CLASSIC GEOLOGY IN EUROPE 7

Cyprus

T0256431

CLASSIC GEOLOGY IN EUROPE SERIES

For details of these and other earth sciences
titles from Dunedin Academic Press
see www.dunedinacademicpress.co.uk

Cyprus

Stephen Edwards
University College London

Karen Hudson-Edwards
Birkbeck, University of London

Joe Cann
University of Leeds

John Malpas
University of Hong Kong

Costas Xenophontos
University of Hong Kong

TERRA

First Published in 2010 by Terra Publishing,
an imprint of Dunedin Academic Press Ltd

Dunedin Academic Press Ltd
Head Office: Hudson House, 8 Albany Street, Edinburgh EH1 3QB
London Office: 352 Cromwell Tower, Barbican, London EC2Y 8NB
www.dunedinacademicpress.co.uk

ISBN 978-1-903544-15-0
© Stephen Edwards, Karen Hudson-Edwards, Joe Cann, John Malpas,
Costas Xenophontos 2010

British Library Cataloguing in Publication Data
A catalogue record for this book is available from the British Library

Typeset by Roger Jones in Palatino and Helvetica
Printed by CPI Group (UK) Ltd., Croydon, CR0 4YY

Contents

Preface

Cyprus is recognized internationally as hosting the type ophiolite, the Troodos massif, a fragment of oceanic crust and upper mantle that records the processes of seafloor spreading and mineralization. Indeed, such is the exposure, preservation and geological importance of the ophiolite, and the rocks that surround it, that many hundreds of geologists – professional and amateur, student and teacher – make a pilgrimage to Cyprus each year. Because it is a truly classic area of geology, which also provides excellent opportunities to investigate applied and environmental earth science, we decided that it was time for a fully comprehensive and contemporary geological guide of this historic Mediterranean island.

All five of us have been influenced by the geology of Cyprus, particularly as we have all studied processes associated with the opening and closing of ocean basins. We have written the guide in order to share our collective knowledge and understanding of the island with as broad an audience as possible in order to stimulate interest in earth science and field investigation. Although primarily a field guide, we trust the book will also serve as a core text on the geology of Cyprus.

Despite the improving access to northern Cyprus, the stops presented all lie within the southern half of the island and the information we have provided on each serves as an introduction. It is up to the individual explorer to examine the exposures in the detail they require and to discover things for themselves. As with all guidebooks, please bear in mind that routes, developments and exposures change.

There is a vast literature on the geology of Cyprus and, in an attempt to produce uncluttered readable text, we have not followed formal bibliographical referencing protocol, but a list of key articles is included. We trust that this will be acceptable to our friends and colleagues who have undertaken much of the work outlined in this guide.

Finally, and most importantly, we hope the guide allows you to share our passion for, and pleasure of, field investigation and discovery.

Stephen Edwards, Karen Hudson-Edwards, Joe Cann, John Malpas,
   Costas Xenophontos
September 2009

Acknowledgements

Our guidebook is the culmination of many years of field-based teaching and research in Cyprus. Over this time we have interacted and worked with a vast number of professionals and students, and much of their work is represented in some way in this guide. It is impossible for us to acknowledge them all, but they know who they are and we thank them wholeheartedly. However, there are certain individuals who deserve special mention. Ian Gass, Eldridge Moores, Alastair Robertson and Bob Wilson have made huge contributions to the geological study of Cyprus. Colleagues at the Cyprus Geological Survey Department in Lefkosia have provided great support over the years. In particular, Ioannis Panayides, Doros Akritas, Zomenia Zomeni and Antonios Charalambides have given us much assistance, providing us with maps and reports, and meeting all of our demands for material for this guidebook. Thanks are also due to the past senior members of the survey, especially George Constantinou, Andreas Panayiotou and George Petrides, for encouraging geological studies in Cyprus. The information on the Skouriotissa mine would not have been complete without the assistance of Hellenic Copper Mines Ltd, particularly Nicos Adamides and his excellent tours. Logistics within Cyprus have been easy over the years thanks to George Kapnos and his fellow coach drivers, who have ferried innumerable parties of geologists around the island – they even know most of the stops. The people of Cyprus have always made us very welcome, and special thanks for their hospitality are due to Carolyn Elliott-Xenophontos, Eleni and Andreas Xenophontos at the Troodos campsite, and Andreas and Yvonne Georghiou at the Axiothea hotel in Pafos. The initial work on this guide was undertaken while Stephen Edwards was at the University of Greenwich, and the universities of Greenwich and Hong Kong are acknowledged for financially supporting some of the fieldwork in Cyprus.

Other individuals made specific contributions to the guide. Carolyn Elliott-Xenophontos, Charlie Underwood and Geoff Parsons kindly read and improved parts of the text, and John Hall, Crispin Little and Charlie Underwood generously supplied some of the figures; Gavin Chan drafted several of the diagrams. Roger Jones, our publisher, demonstrated great efficiency and patience throughout the preparation of the guide. Last, but by no means least, we sincerely thank our families, friends and work colleagues who have supported our work in many different ways and who have eagerly awaited the completion of this project.

The following kindly permitted us to modify or adapt items originated by them:

Blackwell Publishing: Figure 6.1

Christopher Harrison: Figure 2.1

Cyprus Geological Survey Department: Figures 1.2, 1.4, 2.2, 2.5, 3.3, 5.1, 5.2, 5.6, 6.1, 7.7; Table 6.1

Cyprus Water Development Department: Figure 7.9

Geological Association of Canada: Figure 2.1

Geological Society of America: Figure 2.2

Geological Society of London: Figure 5.3

Geological Survey of Israel: Figure 1.2

Istituto Nazionale di Oceanografia e di Geofisica Sperimentale: Figure 1.1

Macmillan Publishers Ltd (*Nature, Physical Science*): Figure 2.4

Introduction

Cyprus: an island of outstanding geology and history

"Welcome to Cyprus!" is a greeting heard everywhere on this Mediterranean island. It reflects the warmth Cypriots have towards visitors coming to experience the bounties of their island – rich in history, romance, folklore, hospitality, fresh produce, sunshine, scenery and, of course, geology. Perhaps nowhere else on Earth does so small an area provide such an excellent illustration of Earth processes through abundant exposures of spectacular and diverse geology. This superb and accessible natural laboratory records at least 200 million years of plate-tectonic activity, particularly the formation and uplift of parts of the sea floor.

To study the geology of Cyprus is also to learn about human history and the development of Western civilization. Situated in the eastern Mediterranean, the island is in a position that has proved to be of repeated strategic importance throughout historical times, as it lay at the centre of the ancient world, at the crossroads between Europe, Africa and Asia. In the development of their cultures and civilizations, the ancients gained an understanding of geology and geomorphology, and were able to exploit and gain wealth from the natural resources of the island, such as copper minerals, building stone, soil and water. There is no question that the geology (both pure and applied), geomorphology and archaeology of Cyprus are truly classic and of world class. They deserve promotion and preservation for their educational, scientific and recreational value – hence this field guide.

Using the guide

The guide is written with the well informed amateur and undergraduate geologist in mind, and uses a thematic approach that logically and comprehensively covers most aspects of the geology of Cyprus. Following on from Chapter 1, as an introduction to the island, Chapters 2 to 7 each focus on a specific theme in which the stops are described in a stand-alone

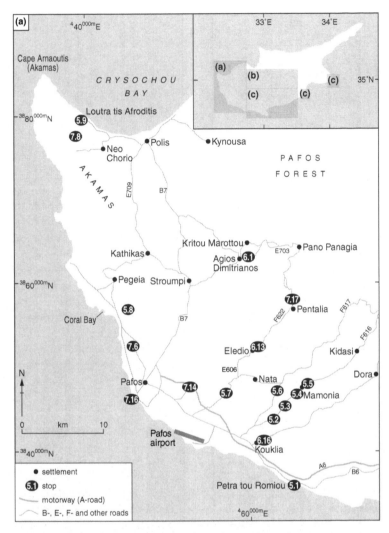

Figure I.1 (above and following) Location of stops. The prefix to each stop number corresponds to the chapter in which the stop is described. Border numbers refer to the WGS84 UTM grid.

fashion. This format enables users to select stops that suit their needs and available time, but, for those who prefer pre-defined itineraries, there are some day-long excursions described in Chapter 8.

The stops within a theme may be scattered over a large area, but a prefix to the stop number clearly identifies the chapter and theme (Fig. I.1). Directions to each stop are usually given from a starting point on a main road or with respect to the nearest town or village. Each stop contains at least

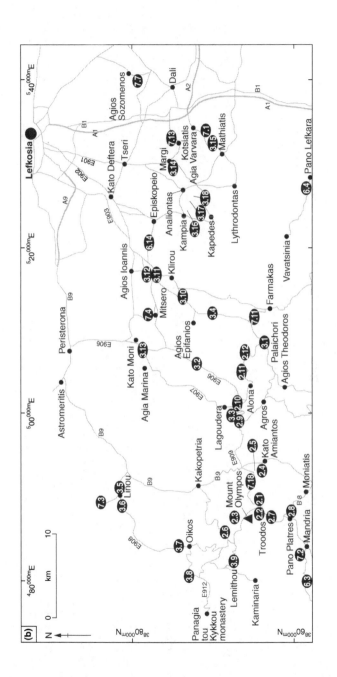

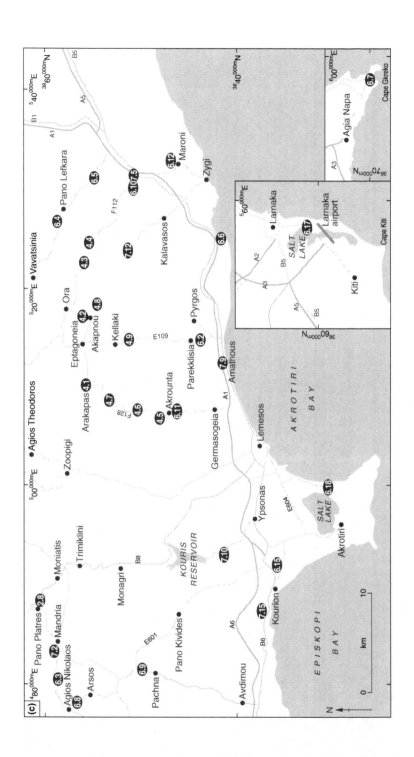

one coordinate that corresponds to a position determined in the field with a global positioning system (GPS) using the World Geodetic System 1984 (WGS84) horizontal datum. All coordinates are within zone 36 of the Universal Transverse Mercator (UTM) Grid, but elevations are with respect to mean sea level and have been taken from published maps.

For most users of the guide there will be no need to acquire topographical maps, but for those who do, the K717 series 1:50000 land maps of Cyprus (1999–2000) use WGS84, but many other maps, such as the DLS 17 series 1:5000 topographical maps (1976), use the European 1950 (ED50) datum. To convert WGS84 to ED50 coordinates, add 27 m to the easting and 179 m to the northing. When using a map and a compass in Cyprus, bear in mind that in 2003 magnetic north was of the order of 3° east of grid north and increasing each year by 1′ east.

All geographical names in the guide follow the standardization agreed by the Cyprus Permanent Committee for the Standardization of Geographical Names in 1991, which are presented on the 1:250000 *Survey of Cyprus administration and road map* (2001). Although these names may vary from those seen on other maps and on roadsigns, the difference is usually subtle and will not cause any problems for navigation. For example, "y" is being replaced by "g", as in Agios (formerly Ayios) and Germasogeia (previously Yermasoyia). Perhaps less obvious are Lefkosia, Lemesos and Choirokoitia, which were formerly Nicosia, Limassol and Khirokitia, respectively. All geological names follow those on the 1:250000 *Geological map of Cyprus* (1995) produced by the Cyprus Geological Survey Department.

Orientation and getting around

The guide is written assuming that all readers will be equipped with a good road map of the island and the 1:250000 *Geological map of Cyprus* (1995). The 1:250000 *Mineral resources map of Cyprus* (1982) and 1:250000 *Hydrogeological map of Cyprus* (1970) are also recommended. These three geological maps may be obtained in Lefkosia from the Cyprus Geological Survey Department, which also produces memoirs, bulletins, books and other maps of the geology of the island. For general tourist information about Cyprus, a visitor guide is recommended, such as one of those published by Lonely Planet or Rough Guides. Equally, the Cyprus Tourism Organisation produces informative maps, booklets and pamphlets that are free of charge, and these may be obtained from its offices abroad or on the island, in the main towns and cities and at Larnaka and Pafos airports.

Large parties visiting the island will usually use a coach with a driver who will know the roads and possibly the stops. Individuals and small groups are better off hiring a vehicle, as this permits the greatest flexibility and many stops are off the well beaten track. However, bear in mind that most rental vehicles, even of the four-wheel-drive variety, are not insured for use on unpaved roads. It is advisable to check a rental vehicle for road worthiness and an adequate spare wheel before driving away for the first time, and remember to drive on the left in Cyprus. When navigating, be mindful that no guide is immune to changes in directions and this is no more true than in Cyprus, where roads and routes are changing all the time, especially in and around towns and villages and in the mountains. Do not be concerned if F and E prefixes to road numbers vary; the key is the number. Rushing about in Cyprus is usually futile, as it is not worth competing against the hospitality, topography and climate, so adjust accordingly and enjoy the way of life, and also allow time to visit several of the superb archaeological sites and monasteries.

Field health, safety and equipment

Health and safety in the field cannot be covered fully here, but they are of course generally a matter of common sense, and of respect for oneself, others and the environment. Details on field safety and codes of practice in the field may be obtained from the Geological Society of London and the Geologists' Association in London. An informative interactive module on field safety for geologists is available from the UK Earth Science Courseware Consortium.[*]

Before departing for Cyprus, make sure that inoculations are up to date, adequate insurance is taken out, and travel details, plans and contact details are left with an appropriate person. As a courtesy and a safety measure, anyone intending to undertake geological fieldwork of any kind in Cyprus is advised to inform the director of the Cyprus Geological Survey Department well in advance of a visit. Should an emergency arise in Cyprus, remember that most people understand at least some English and the emergency telephone numbers are 199 or 112.

Always dress appropriately for the terrain and climate, and carry an emergency medical kit. Sturdy ankle-high boots are recommended whatever the conditions, especially as one of the most common field injuries is a twisted or sprained ankle. A long stick is a good companion when working off road and may be very useful for warding off inquisitive animals and warning others, especially snakes. It is prudent to wear a high-visibility

---

[*] Details at www.ukescc.co.uk.

top, especially when examining road exposures and during the hunting season, and a safety helmet should be worn whenever working at cliff faces.

In winter the mountains may be very cold, wet, icy and misty, in stark contrast with the extreme heat in the lower lands in the summer. It is generally sunny and hot all over the island in summer, so be cautious about exposure to sunshine and wear a hat and sunglasses (the latter are particularly necessary when examining white rocks). Although rare, dust storms may occur, particularly in the spring, and these can cause serious difficulties for people with respiratory disorders, who should wear a mask or stay indoors during these events.

Always carry plenty of water in the field. Most tapwater seems to be safe to drink, and bottled water is sold everywhere. Do not hesitate to drink from the plentiful spring-fed fountains in the Troodos mountains. Fortunately, cafés, tavernas and restaurants are abundant and there should be no problem obtaining food and refreshment during a day in the field. However, during the winter months, facilities in many of the mountain villages are closed.

In terms of specific equipment it is advisable to carry a penknife, hand lens, compass and GPS, the latter two invaluable for following the directions in this guide. Do not collect geological specimens of any kind unless there is a very good reason to do so, and under no circumstances remove archaeological materials, as they are protected by law. Consequently, a hammer and protective glasses are not really necessary. If sampling is to be undertaken for a specific reason then it is advisable to seek prior written permission from the Cyprus Geological Survey Department. Most geologists like to take photographs in the field, but think before shooting, because it is forbidden to take photographs of military establishments and of the buffer zone separating the northern and southern parts of Cyprus. These are not always immediately obvious and may well be in the far distance. The police and military will detain individuals who arouse suspicion, so carry identification at all times and, if available, a letter from the director of the Cyprus Geological Survey Department, which authorizes fieldwork.

Chapter 1

Cyprus: environment and history

Location and size

Cyprus is situated in the northeastern corner of the eastern Mediterranean basin and is flanked by the continents of Europe, Asia and Africa (Fig. 1.1). The island has a maximum length of 225 km, from westernmost to easternmost tips, and a maximum width of 94 km. Occupying an area of 9251 km^2, with 853 km of coastline, Cyprus is the third largest island in the Mediterranean Sea after Sicily and Sardinia. For such a small area, the geology is spectacular, well exposed and highly accessible.

Geological history

The closure of the Tethys Ocean

The Mediterranean Sea represents the last vestige of the once much larger basin of the Tethys Ocean, which dominated global geography in Mesozoic time. Plate movements have gradually closed this ocean over the past 100 million years, and closure will be complete in a few million years' time, when the African continent finally collides fully with the continent of Europe. The end product will be a mountain range much like that of the Himalayas to the east, where the continents of India and Asia have already collided following the complete closure of the Tethys in that region. It is ironic that the plate-tectonic processes responsible for the formation of Cyprus will eventually destroy it and incorporate it into a huge mountain belt.

The tectonics of the eastern Mediterranean are complex, with microplates jostled around in the collision zone between the much larger African and Eurasian plates. Sitting within this broad zone of plate convergence, Cyprus remains tectonically unstable and seismically active. The island lies to the north of the Cyprus trench (Fig. 1.1), which is the southern

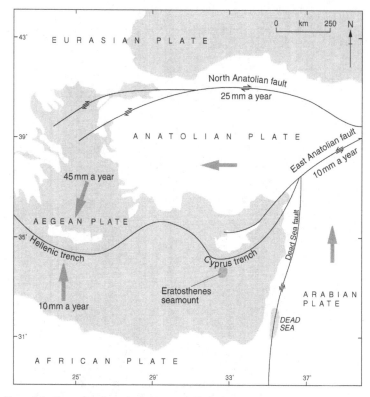

Figure 1.1 Present-day tectonics in the eastern Mediterranean region. Arrows indicate motion relative to the Eurasian plate. Modified from Papazachos et al. (1998).

expression of an east–west-striking subduction zone that dips to the north under Cyprus (the Cyprean arc), and has been under the influence of this zone for at least the past 10–20 million years. For 180 million years prior to this, the geological evolution of the island was likewise dominated by plate interactions, whether these involved the generation or destruction of oceanic lithosphere, the accretion of oceanic seamounts, or the collision of micro-continental blocks.

The remnants of plate interaction are preserved in Cyprus as three major terranes – Troodos, Mamonia and Kyrenia – that give rise to distinctive relief, but also lie beneath a series of cover rocks known as the circum-Troodos sedimentary succession (Fig. 1.2). These terranes represent pieces of lithosphere that have been brought together by tectonic processes, and the complex and varied geology that this has produced in Cyprus is repeated in neighbouring countries. For example, Troodos-like terranes have been identified from Greece in the west to Oman in the southeast, and

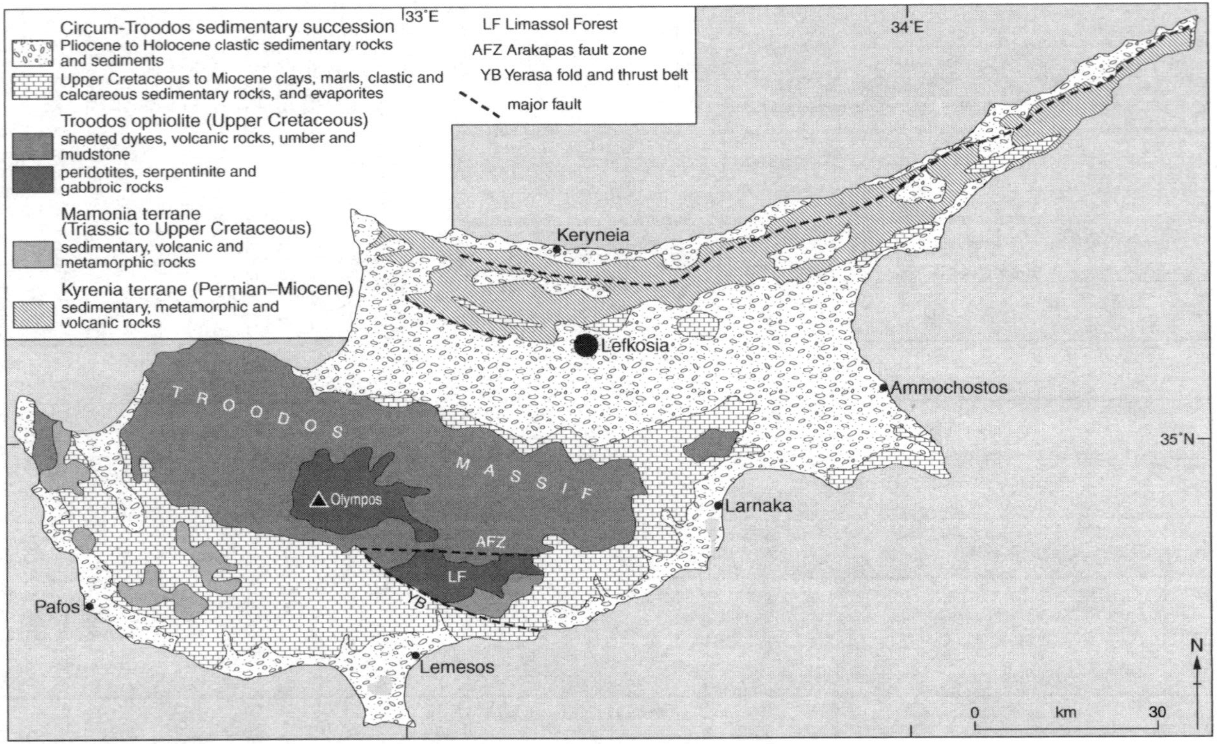

Figure 1.2 (a) Geology of Cyprus, after Cyprus Geological Survey Department (1995). (b) Digital shaded relief map of Cyprus, kindly prepared and supplied by John K. Hall, Geological Survey of Israel.

(a)

Circum-Troodos sedimentary succession
  Pliocene to Holocene clastic sedimentary rocks and sediments
  Upper Cretaceous to Miocene clays, marls, clastic and calcareous sedimentary rocks, and evaporites

Troodos ophiolite (Upper Cretaceous)
  sheeted dykes, volcanic rocks, umber and mudstone
  peridotites, serpentinite and gabbroic rocks

Mamonia terrane (Triassic to Upper Cretaceous)
  sedimentary, volcanic and metamorphic rocks

Kyrenia terrane (Permian–Miocene)
  sedimentary, metamorphic and volcanic rocks

LF Limassol Forest
AFZ Arakapas fault zone
YB Yerasa fold and thrust belt
--- major fault

Keryneia
Lefkosia
Ammochostos
Larnaka
Lemesos
Pafos
△ Olympos

TROODOS MASSIF
AFZ
LF
YB

33°E 34°E
35°N
0   30 km
N

they formed at the same time as the Troodos terrane, about 90 million years ago in the Upper Cretaceous.

The Troodos terrane

The Troodos terrane is considered to represent part of the Troodos microplate and it comprises a complete and relatively undeformed ophiolite sequence that preserves oceanic lithosphere formed during a period of seafloor spreading above a subduction zone (Fig. 1.3). Since its formation, the Troodos ophiolite had rotated 90° anti-clockwise prior to the Lower to Middle Eocene, and it has moved northwards by 10–15°. It has also been updomed to produce an oval outcrop pattern of ultramafic rocks in the centre, which are surrounded successively by gabbros, sheeted dykes, and volcanic and sedimentary rocks (Fig. 1.2a). This pattern is well developed in the Troodos massif, but somewhat disturbed in the southeastern portion of the ophiolite along the Arakapas Valley, which may represent the surface expression of a fossil transform fault, and, to its south, the structurally complex area referred to as the Limassol Forest. The ophiolite is important economically because it is mineralized with copper-bearing sulphide deposits in the volcanic rocks and chromite and asbestos in the ultramafic rocks.

The Mamonia terrane

Compared with the huge bulk of ophiolitic rocks in the Troodos massif and Limassol Forest, rocks of the Mamonia terrane crop out within relatively small and isolated erosional windows in the southwest of Cyprus (Fig. 1.2a). The Mamonia rocks consist of a colourful, jumbled and deformed mixture of Triassic lavas, Mesozoic sedimentary rocks and subordinate metamorphic rocks, but associated with them are fault zones along which slivers of Troodos serpentinite and volcanic rocks occur. The Mamonia terrane probably records the break-up of a Mesozoic passive continental margin during a period of plate convergence with the Troodos microplate (Fig. 1.3). The Mamonia and Troodos rocks were brought together in the Upper Cretaceous through the continuing northward subduction of the Tethyan oceanic lithosphere that lay to the north of the Mamonia margin, a process that had already resulted in the formation of the lithosphere of the Troodos ophiolite during a period of seafloor spreading above the subducted slab. The contact between the two terranes is, therefore, a complex tectonic one.

The Kyrenia terrane

The Kyrenia terrane stretches east–west along the entire length of the mountainous northernmost part of Cyprus, the Kyrenia Range (Fig. 1.2),

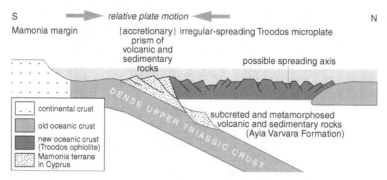

Figure 1.3 Convergence between the Mamonia margin and the Troodos microplate in the Upper Cretaceous.

and has structures reminiscent of a small Alpine-type fold and thrust belt. The terrane has a long and complex history and it contains the oldest rocks in Cyprus, which are shallow-water limestones of Permian age that occur as blocks in large-scale debris flows of Eocene age. These blocks are thought to be remnants of a carbonate bank that originated on the northern margin of Neotethys during its initial formation after the break-up of the supercontinent Pangaea. The carbonates continued to develop throughout the Triassic and Jurassic, but were tectonically disturbed and in places metamorphosed to marbles and schists in the Cretaceous, at the same time as the Troodos ophiolite was forming to the south. The platform carbonates and their metamorphic equivalents are overlain by less deformed successions of volcaniclastic sedimentary rocks, pelagic carbonates and volcanic rocks. The volcanic rocks and associated sedimentary rocks are similar in age and composition to those that lie immediately above the ophiolite in southwest Cyprus and may have formed in an island-arc environment. Overlying these rocks is a succession of pelagic carbonates with interspersed volcanic rocks that appear to have an ocean-island affinity. These within-plate volcanics suggest that the area was undergoing extension at this time (Maastrichtian to Upper Palaeocene), possibly related to the rotation of the Troodos microplate to the south.

During the Upper Eocene the Kyrenia rocks were severely disturbed and deformed by a period of south-directed thrusting that uplifted and exposed lithologies to erosion, with the consequent formation of large-scale debris flows and landslides (olistostromes) that poured off the advancing thrust sheets. Locally, some lithologies were metamorphosed, and schists and marbles formed. The deformed basement lithologies of the Kyrenia terrane are overlain by clastic sedimentary deposits, mostly of local derivation. Further uplift in Plio-Pleistocene time, along with the Troodos massif to the south, tilted the Kyrenia stratigraphy to steep angles. This

13

uplift was again a result of compressional tectonics associated with the subduction zone to the south and, more locally, with underthrusting of the Troodos terrane beneath the southern edge of the Kyrenia terrane.

The circum-Troodos sedimentary succession and the uplift of Cyprus

The circum-Troodos sedimentary succession onlaps and covers rocks of the Troodos, Mamonia and Kyrenia terranes. This succession is Upper Cretaceous to Holocene in age and it passes upwards from hydrothermal and deepwater sedimentary rocks belonging to the Troodos ophiolite, through clay-rich lithologies, calcareous sedimentary rocks and evaporite deposits, to clastic sedimentary rocks and sediments at the top. The older sedimentary rocks are best developed to the south and east of the Troodos massif, whereas the younger clastic rocks and sediments are most extensive between the mountainous areas defined by the Troodos and Kyrenia terranes (Fig. 1.2). The deposition of the sedimentary cover reflects changes in tectonic activity and relative sea level that have taken place in a mainly compressional tectonic regime that has existed since the collision of the Troodos and Mamonia terranes in the Upper Cretaceous. The continuing effects of plate convergence and rotation resulted in pulses of compression and extension, with the development of local depositional basins and isolated highs, and the rapid uplift and erosion of the Troodos massif, which peaked in the Pleistocene.

Many parts of Cyprus began to emerge from the Mediterranean Sea during the Miocene, especially areas of the Kyrenia Range and Troodos massif, but it was not until about a million years ago, during the Pleistocene, that most of what is now the island of Cyprus was above sea level. The uplift continues to this day and is driven by the tectonic underthrusting of the continental crust of the African plate as a consequence of northward subduction beneath Cyprus. Underthrusting, currently of the Eratosthenes seamount south of the island, has produced uplift over a large area, including the Kyrenia Range and perhaps even the Misis Mountains in southern Turkey. More local uplift centred beneath Mount Olympos in the Troodos massif has resulted from subduction-driven serpentinite diapirism. The continued occurrence of earthquakes in and around Cyprus indicates that subduction is still taking place beneath the southern part of the island and that collision between Africa and Europe is continuing.

Landscape and topography

The geology of Cyprus has a major influence on the landscape and topography of the island (Fig. 1.2). As Cyprus was not glaciated during the most recent ice age, the landscape has been shaped mainly by tectonic upheaval, extensive erosion and mass movement, and deep incision by rivers. The highest ground is mostly occupied by rocks of the Troodos, Mamonia and Kyrenia terranes, and the lower ground is dominated by cover rocks of the circum-Troodos sedimentary succession.

The highest point on the island is Mount Olympos in the centre of the oval-domed bulk of the Troodos massif. At 1951 m above sea level, it towers above the 1024 m peak of Kyparissovouno, the highest point in the Kyrenia Range, whose hills and mountains form the narrow east–west-trending spine along the northern limit of the island. These mountain ranges and their foot-hills face one another across a central plain that lies below the 200 m contour. This plain and the widening plains to the west and east make up the Mesaoria (literally meaning "between the mountains") Plain, the western extent of which is sometimes referred to as the Plain of Morfou. Many other plains exist, but these are narrow strips of land sandwiched between the shoreline and the mountains. They are most abundant along the southern and eastern coasts, where there are also long sandy beaches in many bays and coves, and occasional salt lakes. By contrast, the western and northern coasts are more indented and rocky, owing to the mountainous and hilly terrain in these areas.

Climate and meteorology

The seasonal rhythms of hot, dry and often humid summers, and changeable wetter winters, are typical of the intense Mediterranean climate experienced by Cyprus. Summer extends from mid-May to September or October, and winter from mid-November to mid-March. Spring and autumn are therefore short and are seasons of rapid changes in weather conditions. The frequently cloudless skies and long days in the summer months give rise to mean daily temperatures in July and August of 29°C on the Mesaoria Plain and 22°C in the semi-alpine Troodos Mountains. This may not seem very hot, but temperatures of over 40°C are known in Lefkosia and in sheltered mine pits and river canyons at low elevations. Winters are mild, with mean temperatures of 10°C on the central plain and 3°C on the highest parts of the Troodos.

The mountains of Cyprus are responsible for a microclimate over the island producing an average of about 500 mm of precipitation annually.

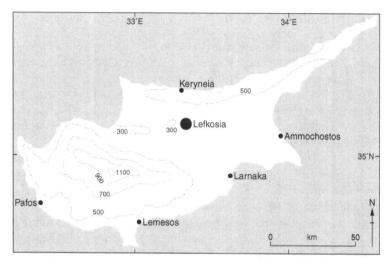

Figure 1.4 Map of annual precipitation (in millimetres) over Cyprus during the first half of the twentieth century (modified from Cyprus Geological Survey Department 1970).

This means that Cyprus, like Lebanon, is relatively green and lush, rather than being a semi-desert only a few hundred kilometres north of the Sahara. The winds in Cyprus are normally light to moderate, and broadly come from the west (from northwest around to southwest). They carry significant moisture, but the amounts of precipitation over the island are highly variable because of the influence of relief, particularly that of the Troodos (Fig. 1.4). Precipitation is highest (400–1150 mm annually) on the southwestern windward slopes of these mountains and on Mount Olympos, and much lower on the Mesaoria Plain to the northeast, where only 300–350 mm falls annually. At least 60 per cent of the average annual precipitation falls during December, January and February. Some of this falls as snow above 1000 m on the Troodos from early December to mid-April, when it may accumulate to considerable thicknesses on northern slopes.

Thunderstorms may occur at any time of the year, but are most frequent from October to May. These storms are usually short lived, but often involve enough precipitation to cause mass movement on the higher slopes of the Troodos and within the clay-rich lithologies of southwest Cyprus. Dust storms are rare, but in spring it is not uncommon for the atmosphere to become filled with yellowish dust brought by strong winds (mainly the Khamsin) from the Arabian and African deserts.

Soils and agriculture

The soils of Cyprus reflect its geology, geomorphology and climate, and give colour to the scenery. Perhaps the most distinctive are the red and orange-brown terra rossas that develop on hard limestones, particularly in southeastern Cyprus, where they are deep and highly fertile, and are used for growing potatoes. In the south and southwest, the decomposition of limestones and marls gives rise to white lime-rich soils that are extensively cultivated, with vines particularly common in the districts of Pafos and Lemesos. Most of the broad river valleys are cultivated, because they host fertile alluvial soils, and groundwater is readily available. Soils on calcareous rocks often contain hard layers of caliche, which are undesirable because they reduce soil quality and subsurface drainage.

On the Troodos ophiolite, the volcanic rocks are overlain by a characteristic brown soil; weathering of sheeted dykes produces a reddish-brown cover dominated by unconsolidated rock fragments. Higher on the Troodos massif, soils are usually thin and sandy, particularly on gabbro, but many slopes are too steep to support a soil cover at all. Partly weathered gabbros become friable and may retain water, so that even up to 900 m or so they are often fertile and highly cultivated. The sparse depleted soils on the ultramafic rocks are able to support only a few species of trees and shrubs that can tolerate the unusual chemistry of these soils.

Agriculture is a major industry in Cyprus and for this reason intense use is made of fertile land almost everywhere. The hillslopes reflect this very well, with their extensive terraces, some dating back hundreds if not thousands of years. The cultivated areas occupy many of the lower hills, parts of the coastal plains and the Mesaoria Plain. The northern coastal plain is particularly noted for its olive and carob trees, and deciduous fruit trees are grown in many of the island's fertile mountain valleys. Vineyards are abundant on the southern and western slopes of the Troodos, where vines can grow at elevations of up to 1000 m in the mild climate. However, it is the Mesaoria Plain that is most extensively cultivated because of its high fertility and flat relief. This and other lowland areas yield vegetables, fruits, olives, carobs, almonds, cereals and fodder crops.

The intensive cultivation of soil involves the application of fertilizers, pesticides and herbicides, and these may eventually find their way into soil, sediment and water. This is not the only impact of agriculture, as many of the cultivated areas receive relatively little rainfall and require extensive irrigation, an activity that accounts for 78 per cent of total water consumption in Cyprus. In order to cope with this demand and to supply domestic water, most of the large rivers draining the Troodos have been dammed and, as a consequence, there are no perennial rivers flowing to the sea. This means that many essential nutrients do not reach the coastal waters around Cyprus.

Flora and fauna

Despite its small size and the impacts of agriculture and hunting, Cyprus hosts richly diverse flora and fauna. Its isolation has led to the evolution of a host of endemic species, but also there are many biological elements characteristic of the surrounding continents. Add to this the variations in topography, climate and soil type that yield a wide variety of habitats, and it is easy to understand why the rich diversity exists. The main habitats of the island may be classified as coastal (including wetlands), low hillsides below 1000 m, mountains above 1000 m, and cultivated land.

Of the 1800 species and subspecies of flowering plants in Cyprus, 128 are endemic and half of these occur in the Troodos, which, along with the Kyrenia Range, also hosts appreciable remnants of the dense forests that covered most of Cyprus in antiquity. The Troodos are covered with pine, dwarf oak, cypress, cedar, juniper, poplar, alder, maple and plane, and many of these trees at high elevations are tortuously twisted or split as a result of lightning strikes that causes sap within them to boil or explode. Where forest no longer exists and the land is uncultivated, the ground is covered by maquis and garigue. The maquis vegetation grows mainly on siliceous soil and includes plants such as rose laurel, arbutus, myrtle and rosemary. Garigue is a scrub vegetation that grows mainly on calcareous soils and includes gorse, caper, lentisk and thyme. Maquis and garigue are most common on hillsides below 1000 m, but maquis-type forests do fringe parts of the coastline.

The fauna of Cyprus is very varied and there are many species of bird, mammal, amphibian, reptile, insect and mite. For example, there are 357 listed species of bird on the island, and every year Cyprus becomes a vital stop-over for vast numbers of migratory birds on their flights between Europe and Africa. This is not only because of the island's location, but also because of the presence of the two coastal wetlands, both incorporating salt lakes, near Larnaka and Lemesos. Efforts are being made to conserve the wildlife of the island, and the moufflon and turtles (green and loggerhead) are all protected. The moufflon is a rare type of wild sheep that is found only on the island and it is the symbol of the Republic of Cyprus, appearing on its coins and as the corporate logo of Cyprus Airways. Attempts are also being made to raise awareness about the eight endemic species of snake on the island, all of which are most active in the spring and autumn, but are timid. Several are venomous, but it is only the blunt-nose viper, or koufi, that poses a real danger owing to the potency of its bite and venom. It is easily distinguished by a yellow horn-like tail and a blunt diamond-shape head.

Human history

The history of human habitation and development of Cyprus has been a key part of the history of Western civilization and can be divided into some 13 periods stretching from 8000 BC until present, with civilization beginning to flourish during the Neolithic period. Cyprus lay at the centre of the ancient world (Asia Minor, Crete, Egypt, Greece, Mesopotamia and Syria), and its culture and development have been very much influenced by these countries. Because of its wealth of natural resources (minerals, rocks, sediment, water, fertile soils and wood), and its climate, topography and highly strategic position between the continents of Europe, Asia and Africa, Cyprus has been repeatedly invaded, settled and developed. This has resulted in a rich and diverse archaeological domain, but it has also left permanent imprints on the landscape, such as those resulting from extensive deforestation caused by the demand for charcoal to fuel early copper production.

The current population of the entire island is of the order of 800 000, 75 per cent of whom are Greek Orthodox. After the Turkish invasion in 1974, the island was split in two by the establishment of a buffer zone. The northern part of Cyprus, representing 38 per cent of the island, became occupied mainly by Turkish Cypriots, but also by Turkish settlers from the mainland. The land to the south of the buffer zone is inhabited predominantly by Greek Cypriots, but also by foreign-born immigrants and seasonal residents, and there are also two British sovereign bases here. Following the division of the island, Lefkosia remained the capital for both sides. Recent political agreements are allowing those living throughout Cyprus to access both the northern and the southern parts of the island.

The area to the south of the buffer zone has a huge transient population that arises from tourism, which is the island's principal industry that currently receives 2.7 million visitors each year. The government ambitiously plans to increase this to 4–5 million in the future, but it is hard to see how this will be achieved logistically and without huge impact on the environment and resources, especially water. By contrast, the area to the north of the buffer zone receives up to only 45 000 tourists annually.

Chapter 2

The Troodos ophiolite: lower section

Introduction

Ophiolites and oceanic lithosphere

The term "ophiolite" was first used nearly 200 years ago by a Swiss geologist named Brongniart to describe occurrences of serpentinite found associated with gabbros, volcanic rocks and chert in mountain belts. It is thus a term used for an assemblage of rock types, which has also been called the "Steinmann trinity" (serpentine, pillow lava and radiolarian chert). Although the association of these rock types has long been recognized, there has been much discussion about their origin and mode of emplacement. Over the past 40 years the igneous portions of the assemblage have been confirmed as being genetically related to one another, a consistent ophiolite lithostratigraphy has been defined, and ophiolites have been accepted by scientists as representing fragments of uplifted oceanic lithosphere.

The composition and structure of *in situ* oceanic lithosphere have been determined through a variety of geophysical techniques, submersible dives and direct sampling methods such as dredging and drilling. These have revealed horizontal layering that is remarkably consistent throughout the oceans and which corresponds specifically to the velocities at which seismic waves travel through rocks of different densities. There is a striking correlation between the seismic layering of *in situ* oceanic lithosphere and the typical ophiolite lithostratigraphy (Fig. 2.1), lending support to the conclusion that ophiolites are indeed fragments of oceanic crust and uppermost mantle.

When examined in detail, however, there are some fundamental differences between ophiolites and *in situ* oceanic lithosphere. As examples, measured thicknesses of rock units in ophiolites often do not correspond to the thicknesses of the seismic layers in the oceans, and there are notable inconsistencies between the mineralogical and chemical compositions of

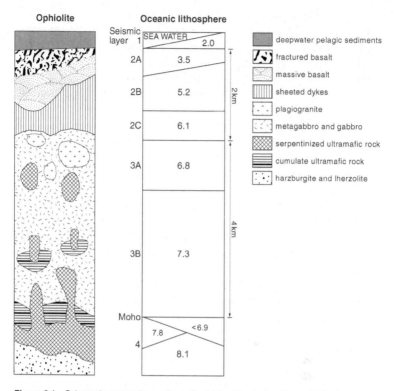

Figure 2.1 Schematic composite sections of ophiolite lithostratigraphy and seismic layers of the oceanic lithosphere (modified from Harrison & Bonatti 1981 and Malpas & Robinson 1997). The increase in P-wave velocity (in km s^{-1}) with depth reflects increasing rock density, the exception being the region yielding velocities of <6.9 km s^{-1}, which corresponds with serpentinized ultramafic rocks. Although the correlation is striking, thicknesses of units in ophiolites do not always correspond with those of seismic layers. The mantle/crust, or layer 3B/4, boundary is called the Mohorovičić discontinuity (or Moho) after the Croatian seismologist who identified it in 1909.

ophiolitic rocks and those of *in situ* oceanic lithosphere. The composition of many ophiolitic lavas is much more like that of island arcs than of mid-ocean ridges, and many mafic and ultramafic rocks in ophiolites and arc environments contain much more pyroxene than their ocean-floor equivalents. It seems likely, therefore, that many, if not all, ophiolites were formed in spreading environments above subduction zones, which is why they are known as supra-subduction zone ophiolites.

The Troodos ophiolite

The Troodos massif and adjacent Limassol Forest are together undoubtedly one of the most remarkable examples of an ophiolite anywhere in the world and, because of superb preservation, exposure and accessibility, one of the most studied. As such, Troodos has played a pre-eminent role in the development of the ophiolite concept and, by analogy, in constraining the processes involved in the formation of oceanic lithosphere. It was in Cyprus, primarily through the recognition of vast expanses of rock made up entirely of dykes (the sheeted dykes), exposed here as nowhere else on Earth, that the connection between ophiolites and *in situ* oceanic lithosphere was first made.

The ophiolite crops out almost entirely within the Troodos massif and Limassol Forest, but there are several small outliers to the west and east. The massif forms an uplifted elongate dome in which the structurally deepest rocks are exposed in the core at the highest point on the island, Mount Olympos (Figs 2.2, 2.3). Mapping by the Cyprus Geological Survey Department defined a consistent lithostratigraphy, from bottom to top, of a mantle sequence dominated by the peridotite harzburgite, a plutonic sequence made up of ultramafic rocks, gabbros and plagiogranites that form the lower crust, a sheeted-dyke sequence, and a volcanic sequence dominated by sheetflows and pillow lavas intruded by dykes (Figs 2.2, 2.4). The contacts between these sequences are either sharply defined structurally or gradational lithologically. The volcanic sequence is capped by the Perapedhi Formation, which is made up of hydrothermal and deep-water marine sedimentary rocks. This formation and the volcanic and sheeted-dyke sequences make up the upper crustal section.

In very simple terms, the formation of the Troodos ophiolite at a submarine spreading centre may be viewed as follows (Fig. 2.4). Upwelling mantle beneath the lithosphere partially melted to produce melt and a solid residue, the latter now represented by the mantle sequence. The melt migrated through the rising mantle to form the crustal rocks. The plutonic sequence crystallized from melt accumulated in a magma chamber within the crust, while sheeted dykes acted as conduits for the transport of melt from the magma chamber to submarine volcanoes, from which lavas were erupted onto the sea floor. This model provides a useful overall framework for studying the ophiolite, but field and laboratory evidence shows that it is too simple. The mantle, plutonic and sheeted-dyke sequences exhibit abundant evidence for successive and multiple injections of melts of variable compositions and the existence of many relatively small magma chambers. The repeated injection and crystallization of melt took place in a dynamic spreading environment, such that rocks formed early in the process were often deformed prior to the injection and crystallization of later melts (Fig. 2.5).

The volcanic sequence is similarly complex. The mineralogical and chemical compositions of lavas are very variable and they suggest that the parent melts must have been derived from different mantle sources,

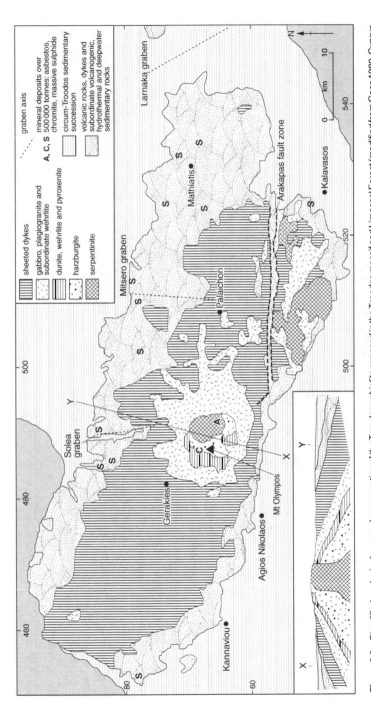

Figure 2.2 Simplified geological map and cross section of the Troodos ophiolite exposed in the Troodos massif and Limassol Forest (modified from Gass 1980, Cyprus Geological Survey Department 1982, 1995 and Varga & Moores 1985). The numbers around the border are from the WGS84 UTM grid, where, for example, easting 480, northing 80 corresponds to the full grid reference 36 480000 E, 38 80000 N.

Figure 2.3 Mount Olympos, looking north from the B8 near Monagri. The mantle sequence crops out on top of the mountain; rocks of the plutonic, sheeted dyke and volcanic sequences are progressively exposed on lower slopes. The foreground is occupied by white carbonate rocks of the younger Lefkara Formation.

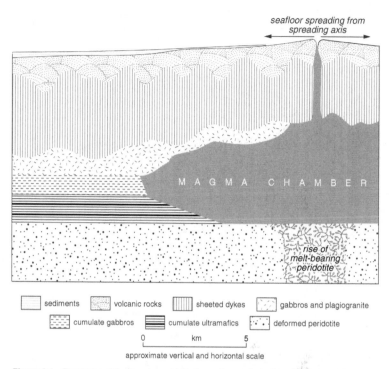

Figure 2.4 represents: seafloor spreading from spreading axis; MAGMA CHAMBER; rise of melt-bearing peridotite.

Legend: sediments; volcanic rocks; sheeted dykes; gabbros and plagiogranite; cumulate gabbros; cumulate ultramafics; deformed peridotite.

0 km 5
approximate vertical and horizontal scale

Figure 2.4 Formation of the Troodos ophiolite from a large, single, long-lived magma chamber feeding a submarine spreading centre (modified from Greenbaum 1972). This simple model defines the major units of the ophiolite: a mantle sequence (deformed peridotites and their serpentinized equivalents), a plutonic sequence (ultramafic and gabbroic cumulate rocks, gabbros and plagiogranites), a sheeted-dyke sequence and a volcanic sequence (mineralized and unmineralized pillow and sheet flows intruded by rare dykes). The ophiolite is capped by hydrothermal and deepwater marine sediments.

perhaps at different times, beneath the spreading centre. This compositional variation within the lavas is also important because it allows conclusions to be drawn about the tectonic environment in which the ophiolite formed. Several types of broadly basaltic to andesitic lava have been identified, and these are virtually indistinguishable from basaltic to andesitic rocks of volcanic islands such as the Marianas and other western Pacific island arcs. Clearly, these lavas are different from those formed at mid-ocean spreading centres such as the Mid-Atlantic Ridge. It is concluded, therefore, that the Troodos ophiolite formed at a spreading centre above a subducting slab of Tethyan oceanic lithosphere (see Fig. 1.3). Magmatism ceased well before a full island arc, similar to the Aleutian Islands, could form. The supra-subduction origin of ophiolitic lavas was first recognized in Cyprus and led to the hypothesis that most ophiolites were formed in very immature island-arc environments.

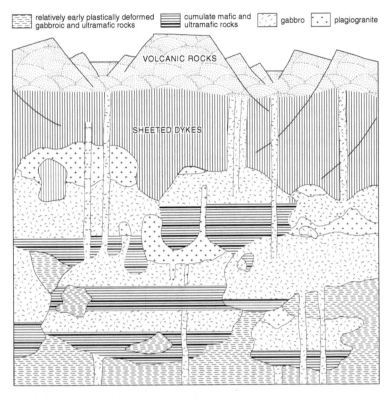

relatively early plastically deformed gabbroic and ultramafic rocks

cumulate mafic and ultramafic rocks

gabbro

plagiogranite

VOLCANIC ROCKS

SHEETED DYKES

Figure 2.5 Conceptual model for the formation of the Troodos ophiolite by multiple injection and crystallization of magma bodies beneath a submarine spreading zone, possibly containing several spreading centres (adapted from Malpas 1990). Spreading-related deformation continued during crystallization of plutonic rocks, so that those formed early are usually more deformed than later ones. Detail omitted for sheeted dykes and volcanic rocks.

Another remarkable feature of the Troodos ophiolite is an obvious lineament that runs east–west across the southern margin of the Troodos massif, separating it from the Limassol Forest to the south (see Figs 1.2, 2.2). This is the Arakapas Valley, which is believed to be the surface expression of a fossil transform fault, akin to those that segment seafloor spreading centres in modern ocean basins. In the Arakapas Valley and the adjacent Limassol Forest, the lithologies are all essentially ophiolitic, but they do not show the regular lithostratigraphy of the region north of the valley. Instead, their interrelationships suggest significant tectonic disturbance and exposure of deeper parts of the oceanic lithosphere on the sea floor, as is common within and close to present-day oceanic transform faults.

The Troodos ophiolite is extensively mineralized (Fig. 2.2). During magmatic activity at the spreading centre, chromite deposits formed in seams and lenses at the top of the mantle sequence and at the very base of the crust. Higher up in the crust, in the sheeted-dyke and volcanic sequences, heat from cooling magma and rock drove hydrothermal circulation of sea water. The hydrothermal fluid vented on the sea floor as black-smoker hot springs and also resulted in the formation of sulphide deposits in the upper crust that contain significant concentrations of copper, zinc and gold. These metalliferous deposits are known as Cyprus-type massive sulphide deposits, and their distribution appears to be fault controlled, as each one is closely associated with a major fault that was apparently active while the deposit was forming. Hydrothermal activity continued well after magmatism had ceased, as sea water, probably derived from oceanic lithosphere subducted beneath the Troodos microplate, entered ultramafic rocks and altered them to serpentinite and, in extreme cases, chrysotile asbestos.

Many ophiolites have a metamorphic sole at their base that was produced during initial displacement of the ophiolite while it was still hot. Such a unit is not obviously present in Cyprus, because the base of the ophiolite is not exposed, but metamorphic rocks of the Mamonia terrane in the southwest of the island have been suggested as candidates.

In order to provide full and manageable coverage of the Troodos ophiolite, it has been broken down in this guide into its major components. The remainder of the present chapter examines the mantle and plutonic sequences, and leads on to Chapter 3, which considers the sheeted dykes, volcanic rocks and sulphide deposits. The region occupied by the Arakapas Valley and Limassol Forest is specifically covered in Chapter 4. A useful place to start for a general introduction to the geology and ecology of a large part of the ophiolite is the Troodos visitor centre (stop 2.1).

Stop 2.1 The Troodos National Forest Park visitor centre   The visitor centre is located in Troodos village on the edge of a large parking area and adjacent to the Troodos tourist pavilion, the Dolphin café–restaurant and tennis courts (36 488865 E, 38 64395 N). It is reached by following the signs for these establishments at the very start of the B8 to Pano Platres.

The centre opened in July 2002 to raise awareness of the park, which was established in 1992 to protect $110 km^2$ of the highest forest in Cyprus. The centre focuses on the geology and biology of the park. There are some useful exhibits and a large topographical model to aid with orientation; a ten-minute video presentation produced by the Forestry Department provides some useful information. Outside there is a 250 m circular botanical and geological trail that is wheelchair friendly and serves as a useful introduction to the plants of the park and the rocks of the ophiolite. There are free trail guides for those intending to hike in the mountains. The centre is open every day (except Christmas Day, New Year's Day and Easter Sunday) from 10 a.m. to 4 p.m. (May to October) and 10 a.m. to 3 p.m. (November to April).

The mantle sequence

There are few places on Earth's surface where rocks from the mantle are exposed. One of these is at the highest point of the Troodos massif, at the heart of this domed structure (Figs 2.2, 2.3). Here, rocks of the uppermost mantle have been brought to the surface by substantial uplift (of the order of 5–10 km) and erosion of the oceanic lithosphere, which has been going on since the Upper Cretaceous. These mantle rocks are dominated by harzburgite and, to a much lesser extent, dunite and lherzolite. Other notable, but very minor, rock types are chromitite, pyroxenite and gabbro.

Harzburgite is composed of olivine, orthopyroxene and minor chromite, but occasionally clinopyroxene occurs and may reach concentrations high enough for the peridotite to be classified as a lherzolite. These minerals are usually deformed and aligned, defining a foliation that was produced by high-temperature plastic flow of the mantle associated with seafloor spreading (Fig. 2.6). On a regional scale, this foliation trends roughly northwest–southeast and dips steeply, which implies that it is primarily a remnant of mantle upflow beneath a spreading centre. However, a component of the dip of the foliation may be associated with much later serpentinite diapirism that caused uplift of the ophiolite. The harzburgite is a residue of extensive partial melting and subsequent efficient melt extraction. In order for the original mantle peridotite to have melted so extensively, water must have been involved to reduce the melting point. Such a situation is often possible in a wedge of mantle above subducted, hydrated, oceanic lithosphere, which lends further support to the argument that the ophiolite formed in a supra-subduction zone environment.

Dunite is composed almost entirely of olivine, but also minor chromite and occasional pyroxene (Fig. 2.6). It occurs throughout the mantle sequence, but is most abundant along the western extent of the harzburgite, where it marks the transition zone from the top of the mantle sequence to the base of the crustal sequence. Dunite in harzburgite occurs as veins, dykes and bodies of less regular shape that are either concordant or discordant with respect to the foliation in the host. The foliation in dunite is defined by grains of chromite and is usually parallel to that in the adjacent harzburgite, suggesting that dunite was formed, and then deformed, in the mantle. Contacts at all scales between harzburgite and dunite may be

Figure 2.6 Weathered surfaces of ultramafic rocks in the mantle sequence (stop 2.3). The rough surface of harzburgite in the lower half of the image shows a vertical foliation defined by pyroxene crystals. In contrast, dunite has a smooth surface and it occupies most of the top half of the image and also cuts the harzburgite. A diagonal vein of coarse-grain pyroxenite cuts these peridotites. The pen is 15 cm long.

either sharp or gradational, and well defined xenoliths and diffuse patches of harzburgite may occur within dunite. Three mechanisms have been proposed for the formation of dunite in the mantle sequence and the mantle–crust transition zone: early crystallization of olivine from intruded melt; melt–harzburgite reaction (whereby pyroxene was dissolved to leave an olivine residue); or, in extreme cases, complete melting out of orthopyroxene from harzburgite.

Pyroxenite, as the name suggests, is mainly made up of pyroxene, and it occurs throughout the mantle sequence and mantle–crust transition zone as bands, veins and dykes, which are either concordant or discordant with respect to the foliation in the host rock. As with dunite, pyroxenite is most abundant in the transition zone, particularly at the base of the plutonic sequence. The origin of the pyroxenite remains puzzling, but it probably formed during fractional crystallization of melt.

Chromite is present throughout all of the peridotites and pyroxenites in the mantle sequence, but is normally a very minor mineral phase disseminated throughout the host rock. Only in a few localities is it concentrated, sometimes forming a rock known as chromitite. These chromitites are generally of crude lozenge shape (known as podiform chromitite), but also occur as layers, veins and dykes. The size, shape, chromite content and texture of these bodies varies greatly, with concentrations of chromite reaching 100 per cent in some massive ores. Where present, the dominant silicate mineral between the chromite grains is olivine or its altered equivalent. Chromitites are always encased in dunite, which exhibits the same features and field relations as the unmineralized dunites described above. The formation of chromitite is, therefore, closely linked to that of its enclosing dunite and it involved a combination of fractional crystallization of melt and melt–harzburgite reaction.

All of the ultramafic rocks exhibit some degree of serpentinization, which is most pronounced in the central and eastern parts of the mantle sequence, where there is extensive shearing, shattering and brecciation. The dominant rock here is bastite serpentinite, formed by the hydration of olivine (to form serpentine) and orthopyroxene (to form bastite) at temperatures up to several hundred degrees Celsius. The bastite serpentinite cropping out in the mantle sequence represents the top of a huge serpentinite diapir that is partially responsible for the upward doming of the ophiolite. A former asbestos mine sits within a zone of extreme alteration in and around Pano Amiantos. Here, veins of whitish fibrous chrysotile asbestos, which formed when fluids migrated through the serpentinite, are associated with pale-green picrolite. Another product of serpentinization is the cream dykes of rodingite that crop out in the area, especially northwest of Kato Amiantos. During serpentinization, calcium and silica are liberated from pyroxene and enter the fluid phase, which then produces rodingite either by direct mineral precipitation or by alteration of pre-existing gabbroic dykes. Serpentinization continues today, enabled by percolating meteoric water and trapped sea water, and is recorded by the hyperalkaline and saline composition of spring waters emerging from the mantle sequence.

Rocks of the mantle sequence within the Troodos massif are restricted to a relatively small area in and around the village of Troodos, which is easy to reach from all the main centres in Cyprus. The village has a range of amenities, including a visitor centre (stop 2.1). In addition to the stops described below, harzburgite and dunite may be examined at stop 2.7, and serpentinite at stop 7.18. Away from Troodos, serpentinite is also encountered at stops 4.6, 4.7, 4.8, 5.6, 5.8, 6.2 and 7.8. For those interested in finding a good variety of chromitite samples in addition to those that can be found at stops 2.2 and 2.3, it is worth investigating the old chromite prospects (36 489883 E, 38 63606 N) 1.5 km to the southeast of Troodos along the dirt road to Mesa Potamos, Vryses and Platres. Wherever mantle rocks are examined, always try to work on weathered surfaces, as fresh surfaces are usually too dark to be informative.

Stop 2.2 The mantle sequence and mantle–crust transition zone along the Atalante nature trail around Mount Olympos The trail is named after the mythological forest nymph, Atalante, and starts by the information board (36 489120 E, 38 64626 N) located in the trees immediately northwest of the postal and telegraph office (an old stone building with green shutters) in the village square in Troodos, very close to the start of the E910 (on some maps the F952) to Mount Olympos and Prodromos. The trail is 9 km long, terminating near the site of the former camp of the Chrome mine on the E910 (stop 2.3), but an extra 3 km can be walked along the E910 or a forest trail to complete a full circuit back to Troodos. The trail circumnavigates Mount Olympos and offers a great day of walking and some superb panoramic views. It enables examination of the geology of the mantle sequence and the lower part of the mantle–crust transition zone, and of the flora and fauna in this part of the Troodos National Forest Park. Points of geological and botanical interest along the trail are identified by numbered posts, and each kilometre is also marked. A very useful free trail guide published by the Forestry Department may be obtained from the Cyprus Tourism Organisation and the Troodos visitor centre (stop 2.1). A brief overview of the geology along the trail is provided here. For those who do not have enough time to walk the trail, the same features can be seen at the other stops in this section.

For the first 2 km the trail passes through harzburgite that is either massive and jointed, or highly shattered, sheared and serpentinized. Orthopyroxene grains define a near-vertical foliation, which becomes a banding where there is variation in the ratio of olivine to orthopyroxene. The harzburgite is cut by veins of dunite, pyroxenite and pegmatitic gabbro. Be aware that during the first kilometre the trail splits close to a V-shape tree; take the lower, well trodden path.

Between 2 km and 3 km the trail continues through harzburgite, but the southwest–northeast-trending ridge on the upslope side of the trail is occupied by a > 400 m-long, late, undeformed, ultramafic–mafic intrusion. This body is highly discordant with respect to the harzburgite foliation and it contains xenoliths of harzburgite in its eastern margins. The intrusion comprises dunite, wehrlite, pyroxenite, gabbro, quartz diorite and dolerite, and is cut by faults trending north-northwest.

Fallen blocks of the intrusion are evident along the trail, with the best around point 22, the sign for Arbutus just behind a tree (36 487962 E, 38 64077 N; it is easy to miss point 22, so head back about 40 m if point 23 is reached. There are many blocks of olivine gabbro and the ultramafic component of the intrusion along the trail. Layering is evident in many blocks, as are crosscutting relationships, such as those exemplified by the late intrusions of pegmatitic gabbro. *In situ* exposures of the intrusion may be examined by traversing up slope in a northerly direction. Layered rocks are well exposed on the ridge crest. From here the whole intrusion may be examined. The best exposures lie along the ridge, and between the ridge and the trail (the southeastern slope). Pyroxenite, gabbro, diorite and dolerite dominate in the west, whereas dunite and wehrlite are more common in the central and eastern parts of the body. The trend of the layering in the west is normally at a high angle to the regional foliation in the harzburgite.

Continuing from point 22, the trail takes a sharp right turn shortly after point 24 and splits into three. Follow the middle trail. From here to the next major bend to the left, where harzburgite is once again encountered, there are boulders of gabbro and dolerite. These boulders are derived from the western end of the late intrusion just examined and may be traced up slope to outcrops of the same rock types.

Between 3 km and 4 km, dunite becomes more abundant and the ratio of olivine to orthopyroxene in harzburgite appears to increase. At point 30 the dunite marked by the sign terminates against harzburgite. It is noticeable that the foliation defined by orthopyroxene in harzburgite continues into the dunite, but in the dunite it is marked by chromite. Another important feature in the dunite is that the concentration of chromite decreases away from the harzburgite/dunite contact. Both of these relationships suggest that the dunite was formed by a process of melt–harzburgite reaction, whereby orthopyroxene was replaced by olivine and chromite.

Just before the 5 km post is point 37, the western adit of the Hadjipavlou chromite mine. Caution must be exercised if the mined area is explored, because there are many unsafe adits and shafts that must not be entered under any circumstances. The main adit is set into a large body of dunite that contains patches of harzburgite, whose foliation is consistent with that of harzburgite hosting the dunite. This again suggests *in situ* replacement of harzburgite by dunite. Examples of chromitite ore may be obtained from the waste pile outside the adit (Fig. 2.7). Just beyond the adit and 5 km marker, the trail narrows where it breaks off to the left; it cuts sharply back on itself where it crosses a stream. The central and eastern parts of the Hadjipavlou mine may be reached by scrambling up the stream bed, in which there are excellent examples of massive chromitite and greyish-green serpentinite. Alternatively, the rest of the mine may be reached by taking the dirt track that crosses the stream and cuts up the slope to the north and east. Exploration of the Hadjipavlou area will show that chromitite ore occurs as a series of discrete bodies, each hosted by dunite in harzburgite. The mineralized zones are deformed and are elongate parallel to the regional harzburgite foliation. The locality is close to the transition between harzburgite and dunite.

Between the signposts for 5 km and 6 km, highly shattered and then

Figure 2.7 Examples of textures of chromite (black) and altered olivine (grey) in Troodos chromitites: **(a)** massive chromite; **(b)** nodular chromite; **(c)** nodular (top left) and disseminated (bottom right) chromite; **(d)** anti-nodular (top left) grading into network (bottom right) chromite.

massive harzburgite gives way to many irregularly shaped bodies of dunite that are hosted by harzburgite. Some of these dunites contain coarse-grain chromite. The significant increase in dunite just before the marker post for 6 km may mark the base of the dunitic mantle–crust transition zone. The abundance of dunite increases further between 6 km and 7 km, and much of this dunite contains patches of harzburgite. Regardless of whether harzburgite is hosting dunite or is contained within it, the concentration of orthopyroxene is noticeably less than it is back along the trail.

At the 7 km marker is an outcrop of massive, internally banded, coarse-grain to pegmatitic olivine pyroxenite and pyroxenite in dunite. The banding is concordant with the regional harzburgite foliation and may have been produced either by repeated injection and crystallization of melt in a conduit parallel to the foliation, or by deformation in the mantle that caused mineral segregation. Between the marker posts for 7 km and 8 km, dunite is continuous, but it is cut by pegmatitic gabbro and pyroxenitic bodies. Some of the latter grade into dunite through a zone of wehrlite. At point 52, a massive dyke of pegmatitic gabbro (with some individual clinopyroxene crystals >15 cm long) has intruded dunite and contains xenoliths of this rock. One margin is in sharp intrusive contact with dunite, but the other is sheared. The gabbro is itself intruded by a dolerite dyke that has well developed chilled margins. Both gabbro and dolerite are near

33

vertical, but are highly discordant with respect to the regional foliation, probably reflecting late magmatic activity.

The trail passes back through the lower part of the transition zone and into harzburgite around the 9 km post. From here onwards the trail is on harzburgite, and soon emerges onto the E910 Troodos–Prodromos road close to the turning to Agios Nikolaos and stop 2.3. From here Troodos is about 3 km away and may be reached by either road or a trail; the latter is joined by taking the dirt road to Agios Nikolaos for about 500 m and then following the white arrow to head eastwards. Both routes back to Troodos pass through harzburgite.

Stop 2.3 Harzburgite, dunite and chromitite at the abandoned Chrome mine north of Mount Olympos   This stop involves a 4 km round-trip to examine the very top of the mantle sequence and base of the mantle–crust transition zone in and around the former Chrome mine on the northern slope of Mount Olympos. The starting and parking point is the turnoff to the dirt road to Agios Nikolaos on the northeast side of the E910 between Troodos and Prodromos (36 487638 E, 38 66 481 N). This turnoff is located about 3 km from Troodos, very close to, and on the same side of the road as, the imposing stone buildings of the camp of the former Chrome mine and nearly opposite the end of the Atalante trail (stop 2.2). The mine is about 2 km away along the dirt road and, although it may be reached using a four-wheel-drive vehicle, it is best to proceed on foot.

As the dirt road gently descends, it cuts through harzburgite and then through a densely forested area. On leaving this forest, where a stream valley develops on the right (eastern) side of the road, the exposure improves and dunite begins to appear more frequently. Stop where the road passes beneath power lines (36 487417 E, 38 67281 N). The exposure on the left of the road is dominated by harzburgite that contains minor concordant bands of dunite, but a large body of dunite lies directly behind it. The next exposure on the left just beyond the clump of trees exhibits superb harzburgite–dunite relationships in addition to some chromite patches and pyroxenite veins (Fig. 2.6). In this exposure, dunite bodies are both concordant and discordant with respect to the foliation in harzburgite, and they occur as veins and irregular patches that have both sharp and diffuse contacts with harzburgite. At least one dunite vein has pyroxenitic margins. There are many crosscutting relationships to observe, implying a period of extended magmatic activity or pulses of melt intruding at different times. Dunite cuts harzburgite, dunite cuts dunite, and pyroxenite veins cut everything.

From here onwards the dirt road passes outcrops of monotonous dunite and two derelict mine buildings, behind which a very pure dunite containing a few minor chromite seams crops out. The view to the north and northeast from this locality is excellent: the Skouriotissa mine (stop 7.3), Morfou Bay, the Plain of Morfou and the Kyrenia Range are all visible. Less than 500 m on from the mine buildings, on a bend to the left, the road splits. Continue straight on along the right-hand track, using the orange and green stripes on a tree for guidance. Follow this track down hill until the derelict buildings of the mine are reached. Be very careful, as there are many potential hazards here: an adit, a shaft, many unstable slopes and buildings, and sharp pieces of wood and metal.

To examine harzburgite, dunite and chromitite, it is best to explore the slopes behind the main collection of derelict mine buildings, clearly identified by the rusty water tank, and those farther to the northeast. The area is dominated by the dunite that forms the lower part of the mantle–crust transition zone. At a few places harzburgite is encountered and the types of harzburgite–dunite relations already observed may once again be examined.

At the base of the slope behind the main collection of derelict buildings is a shaft and a trench cut into a mineralized zone. Massive, anti-nodular and sheared chromitite may be found in this area (Fig. 2.7). Some chromitite bodies are well over 1 m wide. The chromitite is contained within a large body of dunite in contact with harzburgite. On travelling a short distance to the east along the track that passes the adit, more dunite is encountered, in which there are zones of chromitite exhibiting good anti-nodular and network textures (Fig. 2.7). These are best observed in some trenches up slope of the track. During mining, the ore was mapped out as a series of vertical and steeply dipping lenses that were extremely irregular in both shape and size. Their vertical extent was always the longest dimension and their maximum horizontal dimensions generally ran north-northwest–south-southeast, similar to the regional foliation in the adjacent harzburgite. The chromite mineralization in the area is located in the dunitic mantle–crust transition zone. This contrasts with mineralization at stop 2.2 (the Hadjipavlou mine), where chromitite and dunite occur within harzburgite.

The Troodos chromitites were not huge bodies, but the high chromium content of their chromite made the extraction of many of them economic. Chrome mine is one of several mines and prospects dotted across the central and western half of the Troodos mantle sequence. Mining of this area of the ophiolite began in 1924 and continued sporadically until the 1980s. At the Chrome mine, the main activity began in 1934, and in 1936 an aerial ropeway was constructed to transport the ore to a dressing plant near Agios Nikolaos tis Stegis monastery, southwest of Kakopetria. Remains of the ropeway may be observed on the lower part of the mine site, around which good examples of chromitite ore may be seen, especially at the top of the loading area for the ropeway. The ropeway operated until 1953, after which the ore was transported by road.

To return, there are two options: retrace the route in or, preferably, trace the harzburgite/dunite contact by following the power lines up the slope. The latter route will demonstrate the interfingered nature of the contact and the overall regional discordance between the harzburgite foliation, running at a high angle to the north–south-trending power lines, and the harzburgite/dunite contact that runs approximately parallel to these lines. The route will also reveal discordant pyroxenite veins.

Stop 2.4 Serpentinite, rodingite, water resources and unstable mine waste north-west of Kato Amiantos   Kato Amiantos lies to the east of the huge waste piles of the Amiantos mine and is situated along the E801, south of the Karvounas junction of the B9 between Pano Amiantos and Kakopetria, and north of the village of Trimiklini. From the E801 on the northern out-skirts of Kato Amiantos, pull off and park on the dirt road leading off to the west, which starts between the bridge over the Loumata River (imme-diately south) and a war memorial (immediately north).

Some interesting ultramafic and mafic rocks of the plutonic sequence are exposed along the E801 immediately north of the war memorial, but the main objective is to examine the rocks along the Loumata Valley by walking west up the dirt road, keeping the river on the left and being wary of unstable slopes. Along the valley there are exposures of highly sheared and altered gabbro, then ultramafic cumulates and, finally, harzburgite. Many, if not all, of the contacts are faulted, but effectively the Moho has been crossed. After the road becomes a track that abruptly ends, it is nec-essary to continue towards the base of the Amiantos waste pile by walking up the river bed, keeping close to the steep slopes on the right (north). In the boulders there are excellent examples of serpentinized harzburgite cut by veins of serpentine, green picrolite and white fibrous chrysotile asbes-tos (Fig. 2.8). The valley sides expose veins and dykes of cream rodingite and occasional pegmatitic gabbro. Within the rodingites, quartz crystals may been seen and these are evidence that the fluids from which these rocks formed during serpentinization were rich in silica. At the base of the valley side, where it meets the waste pile, is a cave within a near-vertical fault, from which groundwater is discharging. This water has an alkaline composition, which is testimony to reaction between ultramafic rocks and rainwater and snowmelt as they percolate through these fractured rocks.

During the walk back, an appreciation of the problem of instability and transport of mine waste may be gained. The base of the waste pile is the location of many perennial springs and seeps for water that has percolated

Figure 2.8 Vein of chrysotile asbestos in serpentinized harzburgite (stop 2.4). The dark alteration halo bordering the vein is evidence that asbestos-forming fluid reacted with the host rock. The coin is 22 mm across.

through the waste pile. This is good news for local farmers who have a year-round supply of water as a consequence, but bad in terms of slope stability. Steep slopes of unconsolidated material such as those seen here are likely to become more unstable if they contain significant quantities of water. Indeed, views up the slope of the waste pile reveal that it is extensively eroded and unstable. There are deeply gullied sections and scars arising from slope failure, and mudflows have been observed after heavy rain. This mass movement means that sediment is getting into the streams and rivers. In an effort to reduce the sediment load of the water in the past, a series of weirs were built down stream. These are now mainly dysfunctional because they are filled with sediment. During and after periods of heavy rain, torrents of sediment-laden water rush down the river, transporting huge boulders and eroding the river banks and bed. As a consequence, sediment derived from the waste pile has accumulated in the Trimiklini and Kouris reservoirs down stream. The major concern in this area is that continued erosion of the leading edge and toe of the waste pile may promote a large-scale failure of the slope. Further details are provided at stop 7.18 (the Amiantos mine).

The plutonic sequence

The Troodos plutonic sequence is the part of the ophiolite that lies between the mantle sequence and the sheeted dykes. It represents the lower part of the oceanic crust and it comprises a variety of peridotites, pyroxenites, gabbros, diorites and plagiogranites, and rocks that lie compositionally between these lithological types. In very general terms, most of the ultra-mafic rocks lie at the base of the sequence, in or immediately above the dunitic part of the mantle–crust transition zone or its faulted equivalent, and the most evolved rocks (diorites and plagiogranites) occupy the top. This broad and simple ophiolite lithostratigraphy led in the past to sug-gestions that the plutonic sequence may have been derived from a single large fractionating magma chamber (Fig. 2.4), but field relations show that this is not the case. The complex history of accretion of the lower oceanic crust involved the repeated injection, fractionation and crystallization of batches of magma in a spreading zone that may well have contained several small spreading centres (Fig. 2.5). Intrusion was aided by the dila-tion of the oceanic crust associated with spreading, which in turn resulted in brittle failure of the colder upper oceanic crust and plastic deformation of the hotter lower crust.

Field relations permit subdivision of the plutonic sequence into early and late plutons based on crosscutting intrusive relationships and type and extent of deformation (Fig. 2.5). The subdivision is clearer in the lower plutonics, but becomes less distinct in the upper part of the sequence, where the degree of plastic deformation is less pronounced. The lowest rocks comprise dunite, wehrlite and a variety of pyroxenites and gabbros, all of which display banding in places. The lowermost ultramafic rocks probably represent the top of the mantle–crust transition zone, the base of which is dunite in contact with harzburgite. The banding resulted from high-temperature plastic deformation, which is also characterized by strong lineations and foliations, and small isoclinal folds. These features are also seen in the mantle lithologies and they suggest that primary igneous textures, layering and intrusive contacts in most of the mantle sequence and the early plutons were significantly modified by deforma-tion and recrystallization caused by plastic flow associated with mantle upwelling and seafloor extension in the spreading zone.

Many of the deformed rocks of the plutonic sequence, and some parts of the mantle sequence (stop 2.2), are intruded by relatively undeformed plutons that clearly preserve igneous features, although in places they also preserve zones of high-temperature deformation. These plutons are made of plagioclase-bearing peridotites, pyroxenites, gabbros, diorites, plagio-granites and dykes of microgabbro. In the lower part of the plutonic sequence, these plutons are dominated by peridotites and pyroxenites, particularly along a 1–2 km-wide north–south-trending zone to the west of Mount Olympos, but plutons higher in the sequence tend to contain more fractionated rock types. This relationship suggests that magma that fed the higher-level plutons was derived from larger plutons at deeper levels that were undergoing fractional crystallization. Magma in the high-level plutons continued to fractionate, ultimately leading to the crystallization of the final products of magmatic differentiation (magnetite-rich gabbros, diorites and plagiogranites), which generally lie immediately beneath the sheeted dykes.

The uppermost part of the plutonic sequence has a complex contact with the sheeted dykes. In most exposures of this contact, plutonic rocks intrude sheeted dykes, indicating that the construction of the dyke unit was almost complete before the intrusion of plutons ceased. In fewer exposures, screens of plutonic rock are sandwiched between dykes, showing that dykes continued to intrude after the plutons had solidified. In yet other exposures, the dyke and plutonic rocks are separated by a low-angle fault. Where plutons intrude into sheeted dykes, these dykes have been partially to completely metamorphosed or assimilated, which is particularly evident in bodies of plagiogranite that contain xenoliths of sheeted dykes. Hybrid diorites occur where assimilation of mafic xenoliths by evolved magma was so extensive that the xenoliths are barely or no longer recognizable. A variety of hybrid rocks were also produced by magma mixing, during which basaltic magma intruded plagiogranitic magma, in places chilled against it and in others mixed with it.

Some plagiogranites and diorites contain large quantities of the hydrous mineral epidote, which probably indicates introduction of water into the magma by assimilation of hydrothermally altered sheeted dykes. Much of this assimilated water was released by magma as it crystallized and it may have contributed to the hydrothermal fluids that ultimately led to mineralization at or near the sea floor. Pegmatitic gabbros – very eye-catching coarse-grain rocks – are particularly common near diorites and plagiogranites, but they also occur throughout the plutonic sequence and in parts of the mantle sequence. Most appear to have crystallized from water-bearing magmas, but their precise origin is not clear.

Gabbro is by far the most abundant rock type in the plutonic sequence. Its primary mineralogy is plagioclase (40–75%), clinopyroxene (25–60%), orthopyroxene (≤25%) and olivine (≤25%), but amphibole, titanomagnetite and quartz may also be present. Variations in the proportions and textures of primary minerals result in many types of gabbro, the most widespread being two-pyroxene gabbro (plagioclase, clinopyroxene and orthopyroxene) and norite (plagioclase and orthpyroxene). Alteration of the primary minerals leads to further variation in gabbro types.

Stops 2.5 to 2.7 examine rocks of the lowest part of the plutonic sequence, including the transition zone. These contrast with stops 2.9 to 2.12, which mainly examine rocks of the upper part of the plutonic sequence that lie close to the sheeted dykes along ridges far to the east of Mount Olympos. Stops 2.5 and 2.9 to 2.12 may be conveniently combined into an excursion that provides a comprehensive overview of the plutonic sequence. Plutonic rocks can also be seen in the Limassol Forest at stops 4.6 and 4.9.

Stop 2.5 Banded ultramafic rocks of the lowermost plutonic sequence intruded by pegmatitic gabbro and dolerite at Pano Amiantos Pano Amiantos is located along the B9 between Troodos to the west and Kakopetria to the north. The stop involves a short walk along a ridge that rises and runs southeastwards behind two old white-walled buildings (36 493130 E, 38 65254 N), which are located on the outside of a sharp bend on the B9 at the very eastern limit of Pano Amiantos, and lie to the east of the Olympos restaurant and south of a water fountain. The safest place to park is by a small complex of old white buildings on the northern side of the B9, west of the sharp bend. Although not the main objective of this stop, the ridge also provides excellent views of the former asbestos mine (stop 7.18).

The ridge provides exposures of banded and folded ultramafic rocks that have been intruded by relatively undeformed pegmatitic gabbro and dolerite. In general terms, the following rock types are encountered progressively on walking southeastwards along the ridge from the buildings: shattered sandy-colour dunite (pyroxene-bearing in places), olivine pyroxenite, pyroxenite and pegmatitic gabbro; pegmatitic gabbro that intrudes, and contains xenoliths of, olivine pyroxenite; pegmatitic gabbro with occasional doleritic dykes; shattered dunite (pyroxene-bearing in places); folded bands of dunite, wehrlite and olivine pyroxenite (Fig. 2.9).

The banded and folded ultramafic rocks probably represent the transition zone, especially as to the west they are in contact with harzburgites

Figure 2.9 Tightly folded ultramafic rocks in the transition zone (stop 2.5). The coin is 28 mm across.

and serpentinites of the mantle sequence, in which the former asbestos mine sits. In contrast, out of sight along the base of the northeast-facing slope of the ridge, the ultramafic rocks are thrust over deformed gabbros that presumably originally formed higher in the plutonic sequence. The origin of the bands in the ultramafic rocks cannot be determined solely from field relations. However, they may represent igneous layers formed at the base of a magma chamber, dykes that fed higher-level magma chambers, or sills injected laterally along the base of the crust.

From the ridge, look slightly west of north at the roadcut along the B9. The cut contains a faulted contact between dark brown rocks of the mantle sequence on the left and grey horizontally layered gabbros on the right. A little farther down the B9, layered gabbro is cut successively by pegmatitic gabbro, chilled dolerite dykes and shear zones.

Stop 2.6 Intrusive relationships in the lowermost plutonic sequence along the F936 northeast of Prodromos This stop examines two roadcut sections that are approximately 700 m apart along the F936 northeast of Prodromos. The rocks in these sections are representative of the lower part of the plutonic sequence, lying about 2 km northwest of the harzburgite/dunite contact of the transition zone exposed in and around the Chrome mine (stop 2.3). The F936 runs between Kakopetria on the B9 and the E908 between Prodromos and Pedoulas. The first locality lies about 1.3 km from the F936/E908 intersection, and there is ample parking on the outside of a bend marked by a leaning concrete telegraph pole (36 485899 E, 38 68932 N). The road section begins just over 100 m to the southeast, opposite the roadsigns warning of a bend to the left and advising a speed of 40 kph, and is almost continuous to a little beyond the parking area.

The roadcut exposes at least three episodes of magmatic activity recorded by peridotites, pyroxenites and pegmatitic gabbro, which are evidence of the complex multiple intrusive evolution of the lower oceanic crust. In the roadcut opposite the roadsigns, there are alternating bands of dunite, pyroxene-bearing peridotite and pyroxenite (probably orthopyroxenite), none of which appear to contain plagioclase. A foliation within the bands is present in places, lying almost parallel to the banding, and is presumed to be the result of plastic deformation associated with seafloor spreading. As at stop 2.5, this banded section probably represents a component of the transition zone, in which the bands may be deformed ultramafic layers, dykes or sills.

A walk northwest towards the parking area will reveal that the banded rocks are not laterally continuous, as they give way to dunite, and all of these earlier rocks are intruded by dykes of pyroxenite and pegmatitic gabbro. Farther still along the section, the exposure is dominated by the pyroxenite, which preserves a variety of complex fractionation or intrusive relationships, as there appear to be at least two generations of pyroxenite crosscutting one another. It contains huge pyroxene crystals, some reaching several tens of centimetres. Throughout the section so far, the ultramafic rocks are altered: some dunites are highly serpentinized and bleached, and there are zones containing steeply dipping bands of serpentinite that may also host picrolite.

After a break in exposure, on the bend to the left after the leaning telegraph pole, the roadcut exposes a very coarse-grain chromite- and plagioclase-bearing poikilitic wehrlite that appears to be undeformed. Regional studies in the area have demonstrated that the banded ultramafics are intruded by this wehrlite, and that the latter may contain huge xenoliths of the former. The age relationship between the wehrlite and the pyroxenite is not possible to determine from the road exposure. The very last magmatic episode recorded in the road section is represented by the thin chilled dolerite dykes that cut all of the other rock types.

The second locality is reached by travelling about 700 m southwest along the F936, to a place about 600 m before the E936–E908 intersection. There is parking on the right (western) side of the road immediately before the sharp bend to the right above the Platania restaurant. The roadcuts around the outside of this bend expose complex relationships between gabbros, wehrlite and pyroxenite, and the whole section to be examined is marked by the rock-filled gabion baskets that prevent rockfall from reaching the road. The cuttings and slopes are obviously unstable, and great care should be taken while exploring them. The section begins on the first bend after the restaurant in the direction of the E908 (36 485433 E, 38 68452 N), and finishes to the northeast of the parking area where the gabion baskets end (36 485608 E, 38 68574 N). The relationships in the road section are extremely complex and require time to appreciate fully; their description here is somewhat brief.

From the bend to the stream, banded foliated gabbro is intruded by poikilitic wehrlite and clinopyroxene-bearing dunite, and all of these rocks are cut by pegmatitic gabbro. The banding in the gabbro may be igneous layering or its deformed equivalent, and close inspection is required to see if layering features can be distinguished (refer to stop 2.7 for more details). After the stream, wehrlite appears to grade into olivine pyroxenite, pyroxenite and gabbro, including pegmatitic gabbro, and be cut by similar pyroxenites and gabbros. Farther along, there is a large pyroxenitic body and farther still there is more wehrlite cut by gabbro. Some pegmatitic gabbros contain white plagiogranite zones, indicating that they may represent the last stages of fractional crystallization of basaltic magma.

42

The wehrlite/pyroxenite contact is interesting, as wehrlite at the contact is usually richer in pyroxene, and the pyroxene is coarser, than it is farther away from the pyroxenite. In places the contact is marked by olivine pyroxenite. There are at least two ways in which this relationship may have developed. The first is that the pyroxenite formed in channels during melt extraction from the wehrlite, and the second is that pyroxene-forming melt intruded and impregnated dunite, and wehrlite was formed. In both scenarios, pyroxenites represent melt channels, and further evidence for this comes from the gabbroic fractions that some of them contain.

Stop 2.7 Layered gabbroic rocks, harzburgite and dunite along the B8 southwest of Troodos   This stop examines two roadcut sections that are about 800 m apart along the B8 between Troodos and Pano Platres. The first exposes layered gabbroic rocks of the plutonic sequence. These are part of a large block faulted down into rocks of the mantle sequence, which are examined at the second locality. About 1.8 km southwest of Troodos, park on the southeastern side of the B8 by the green house made of corrugated metal (36 487948 E, 38 63176 N), opposite a roadsign for "Troodos and residence". A short distance southwest down the road there is continuous exposure of layered rocks.

The layered rocks are mainly varieties of gabbro, locally rich in olivine, but minor pyroxenite and wehrlite also occur. Layers are 1–40 cm thick, laterally continuous, and defined by variations in mineral proportions and grainsize (Fig. 2.10). These layers may have originally been igneous, most probably produced by the accumulation of early-formed minerals to form stratified layers in the lower part of a magma chamber. However, this original igneous feature has been modified by deformation that generated folds (especially Z-shape folds) of different scales throughout the outcrop and also a strong preferred orientation of minerals within the layers. The deformation probably resulted from high-temperature plastic flow at the base of the plutonic sequence during seafloor spreading. As such, the layered rocks are now strictly metamorphic. Separating original igneous features from their metamorphic overprint is difficult, but the uneven bases of some layers may have been produced by erosion or compaction of the underlying layer during layer formation in the magma chamber.

There are late intrusions into the layered rocks that clearly postdate the deformation attributed to plastic flow. The largest of these is a brown ultramafic body that transgresses the layering at a shallow angle, but the most abundant are dykes and veins of pegmatitic gabbro. These late gabbros are inclined at both low and high angles to the layering, and for short distances are sills parallel to the layering. Careful examination will reveal that some of the wider bodies of pegmatitic gabbro have pyroxene-rich

Figure 2.10 Deformed layered gabbroic rocks of the lower plutonic sequence (stop 2.7). The alternating pale and dark layers reflect variations in mineral proportions.

margins that may have a finer grainsize than their cores. It is not clear whether these margins are chilled or are a product of crystal fractionation or melt–rock reaction.

Now walk down hill in the direction of Pano Platres for about 800 m and stop at the arched bridge on the bend of the old road (36 487290 E, 38 63225 N). During the walk it is obvious that plutonic rocks give way to peridotites of the mantle sequence along a faulted contact. Faulting appears to have mostly or completely removed the transition zone. The faulting occurred after the plastic flow recorded in the layered gabbros and it may have been associated with intrusions of wehrlite in the area.

The section along the old road is dominated by harzburgite, which is best examined approximately 35 m east of the old bridge. On weathered surfaces the pale greenish-grey orthopyroxene and minor black chromite grains stand proud of the light-brown olivine and define a strong foliation, which is nearly vertical and strikes approximately northwest–southeast. Those with a keen eye may be able to identify a lineation defined by aligned grains of orthopyroxene and chromite on the foliation plane. This lineation marks the final direction of plastic mantle flow that also produced the foliation. A detailed search on weathered surfaces should reveal a few bright-green grains of clinopyroxene, a mineral that is rare in Troodos harzburgite.

A little farther west, about 25 m east of the old bridge, is a zone of dunite about 8–10 m wide. Its contact with harzburgite is sharp and it crosscuts the foliation in harzburgite at a very low angle. Within the dunite, the concentration of chromite is low, but generally the grains are more equant and larger than those in the adjacent harzburgite. There are some chromite-rich patches and veins distributed throughout the dunite.

Harzburgite and dunite are both cut by rare 1–2 cm-wide veins of orthopyroxenite that are either concordant or discordant with respect to the foliation. Some of these veins are separated from harzburgite by a thin zone of dunite that is relatively enriched in chromite. Such a relationship may suggest initial dissolution of orthopyroxene, and formation of olivine and chromite by melt–harzburgite reaction, followed by later crystallization of orthopyroxene.

The most interesting intrusion is a microgabbro dyke that crops out on the valley side about 10 m west of the old bridge. It is discordant with respect to the harzburgite foliation it cuts, as it strikes nearly north–south and dips steeply west, but it is internally foliated and there is some segregation into zones dominated by plagioclase and pyroxene (some now amphibole). The internal foliation runs parallel to the dyke margins, which show no evidence of chilling. The dyke contains angular xenoliths of harzburgite, as well as lenses of this peridotite where the dyke splits into interconnecting veins. The morphology of the dyke suggests that melt intruded by fracturing the harzburgite. The internal foliation in the dyke may relate to melt flow, but is more probably a deformation feature produced by stress being focused along the melt conduit in which crystallization was taking place. Most of the gabbro dykes in the Troodos mantle sequence are unlike the one here, as they are less deformed and their grainsize is much coarser.

Stop 2.8 Wehrlite intruding gabbro along the B8 near Pano Platres This stop is located between Pano Platres and Moniatis, at the southern end of a 500 m-long stretch of straight road running due south from the trout farm in Pano Platres. Along the straight there is a parking area in the trees on the western (Platres) side of the road, north of a sharp bend and opposite telegraph pole TA90/12A/9/9/15. The exposure of interest is the very steep rockface on the eastern side of the road, a short distance south and around the bend (36 487969 E, 38 60984 N), and it is best viewed in the afternoon when the light is favourable.

Caution must be exercised while examining the rockface, as the road is busy and rockfalls do occur. Note, for instance, the resurfaced road and destroyed crash barriers as a consequence of a large rockfall in 2003. The safest way to appreciate the exposure is from the opposite side of the road, where it is also possible to examine most of the rock types in the fallen debris.

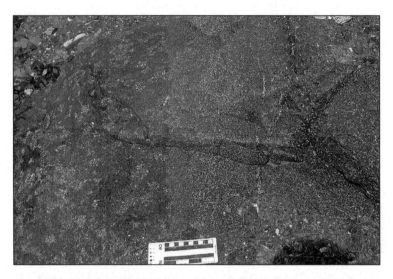

Figure 2.11 Plagioclase wehrlite showing subtle igneous layering (layers run vertically in the photograph) defined by variations in plagioclase (white) content and texture (stop 2.8).

The steep face exposes a complex multiple intrusive history. The earliest rock appears to be a pale gabbro that has been intruded by a darker two-pyroxene gabbro. These have been intruded by a large volume of the darkest rock, a poikilitic plagioclase wehrlite, which in places grades into a dark olivine gabbro at its contact with the earlier gabbros. Huge blocks of gabbro are clearly visible within the wehrlite. The wehrlite/gabbro contact may be sheared and brecciated, and it sometimes hosts dolerite dykes, which represent the last intrusive event. A good spot to examine these relationships is between telegraph poles TA90/12A/9/9/12 and TA90/12A/9/9/11 (36 487969 E, 38 60984 N). At this point it is also possible to see banding in the wehrlite, defined by variations in the content and texture of plagioclase (Fig. 2.11). As the wehrlite exhibits little deformation, this is probably igneous layering or a reflection of late channelling of melt through a crystal mush.

The very sharp, sheared and brecciated contacts between the wehrlite and the gabbro, and the apparent lack of chilled contacts at the edge of the wehrlite, suggest that intrusion of the wehrlite may have been controlled by faults. However, the crude gradation from wehrlite into pale gabbro, through a dark olivine gabbro with, in places, very coarse grain, indicates that some degree of hybridization may have occurred between wehrlite-forming magma and gabbroic host rock.

Stop 2.9 Multiple intrusive events and water resources in the upper plutonic
sequence along the F915 at Chandria   To the east of Kyperounta lies the village of Chandria; the stop investigates roadcuts along a 400 m section of
the F915 on the northern outskirts of Chandria. The section begins about
200–300 m east of the last sharp bend on the F915 as it climbs northwards
out of Chandria and then runs eastwards towards Polystypos. Along this
stretch of road there is parking on a dirt track just east of a sign warning
of bends in the road. The section begins at this roadsign (36 499706 E,
38 66883 N) and continues down hill to the west, around the sharp bend to
the left, and ends along the roadcut covered by netting to prevent rockfall.

The road section provides a superb example of the multiple magmatic
episodes involved in the creation of the upper part of the plutonic sequence.
The plutonic rocks lie close to the contact with the sheeted dykes, which
form the hills immediately above to the north (Fig. 2.12). The extensive
sheets of near-vertical dykes are best seen from the sharp bend. It is worth
spending at least an hour working back and forth along the road section
in order to piece together the complex story of lower crustal accretion
preserved here.

The earliest rocks are dark olivine gabbro, pale gabbro and dark poikilitic wehrlite. These are exposed near the roadsign at the beginning of the
section, but their exact relationship to one another is not entirely clear.
These rocks may be part of a layered unit, or olivine gabbro may be

Figure 2.12 Madari ridge (stop 3.3) viewed from the south, showing the contact (dashed line)
between sheeted dykes (above) and high-level plutonic rocks (below), and the village of Chandria (stop 2.9) below this contact. The ground underlain by plutonic rocks is planted with vines
and fruit trees; in stark contrast, the steep craggy and bare sides of hills are made of sheeted
dykes.

intruded by pale gabbro and then wehrlite. The pale gabbro is the most abundant gabbro along the whole road section and is well exposed at the bend, where it is strongly foliated. It contains coarse pyroxene crystals and occasional black amphiboles set in a medium-grain matrix, and it may exhibit weak mineralogical and grainsize layering. This gabbro does not exhibit chilled margins where it is in contact with earlier rocks. More dark olivine gabbro occurs a short distance down hill from the start of the section, but it appears to be different from that at the start. It exhibits fine-scale layering and, as with the pale gabbro, this may be a product of penetrative deformation.

Farther west along the section, the gabbros have been repeatedly intruded (Fig. 2.13a). A very evolved plagiogranite, consisting mainly of quartz and plagioclase (with secondary green epidote), intruded first. Close inspection of this white rock underlying the gabbros reveals a narrow silica-rich contact zone that contains small partially digested xenoliths of gabbro. Much larger xenoliths occur within the main body of the plagiogranite. Clearly, the plagiogranite represents the top of a pluton that has intruded into gabbros previously crystallized from at least one earlier pluton.

The plagiogranite and gabbros are themselves cut by pale grey-weathering microgabbro and dolerite dykes that show well developed chilled margins (Fig. 2.13a). These dykes are almost vertical and are up to 20 m wide. The latest intrusive event appears to be represented by a series of vertical brown-weathering oxidized mafic dykes that crosscut all other features. These dykes generally have chilled margins and are about 1 m wide. This sequence of intrusive events is also well preserved elsewhere (Fig. 2.13b).

Around the bend is a cut slope covered by netting, and more intrusive relationships are exposed here. The most spectacular are at the beginning of the cut, between the end of a concrete wall and the start of the netting; this point is marked by a telegraph pole (number 86-2) and a drain into a culvert (36 499576 E, 38 66889 N). Here there is a dolerite breccia that formed by fracturing during intrusion of what is now a pale gabbro. The dolerite xenoliths have been metamorphosed, as is evident from the abundance of amphibole, but the timing of this metamorphism is not constrained. The xenoliths also show a significant range of sizes and shapes, both rounded and angular (Fig. 2.13c). Some adjacent dolerite fragments would fit together if the intervening gabbro were removed, but most would not, implying either that they moved within the intruding magma or that their margins were eroded by the magma. Close inspection of the palest gabbroic zones will show that they are composed of plagioclase and green amphibole. These zones become more typically gabbroic where they

Figure 2.13 Intrusive events preserved in the upper plutonic sequence. **(a)** Near-vertical grey mafic dyke (D) cutting white plagiogranite containing xenoliths (X) of dark olivine gabbro (stop 2.9). **(b)** White plagiogranite (P) intruded by an inclined pale-grey microgabbro dyke (D), both of which are cut by three near-vertical dark oxidized dykes (F139 between Zoopigi and Agios Theodoros). **(c)** Xenoliths of grey amphibolitized dolerite in white gabbro (stop 2.9); the coin is 25 mm across.

are charged with many small doleritic fragments, some of which are almost too subtle to recognize. The outcrop provides an excellent example of a gabbro formed by hybridization as evolved basaltic magma fractured, infiltrated and reacted with dolerite. This gabbro is foliated and pale, as seen above the drain, and is similar to the gabbro exposed on the bend and back towards the parking area.

The stop provides an opportunity to consider some important aspects of hydrogeology and water resources. From the parking area or farther east along the F915, the view to the west shows Kyperounta on the relatively flat topography of the gabbros in the foreground, with the steeper terrain of the mantle sequence in the distance. Sitting on the gabbros around the town of Kyperounta are artificial water ponds that are fed by local groundwater. The gabbros are good aquifers because they are fractured and permeable, and their deep weathering provides a sandy overburden that is excellent for recharge. Ponds like those seen here are built to serve the many villages at this elevation in the Troodos, and the ponds and groundwater are monitored closely to make sure that water is not over-used, especially with increases in urban development and agriculture.

Unlike gabbros, sheeted dykes form very steep hills, and soil cover on them is poor or absent (Fig. 2.12). These factors promote runoff and poor water storage. To the east, however, many of the towns and villages are located on, or close to, the gabbro/sheeted-dyke contact, because it is a site of springs. The same can be said of settlements elsewhere in the mountains. For example, Kakopetria (literally meaning "bad rock") to the north of Troodos has abundant springs owing to the presence of fractured ("bad") rocks.

Stop 2.10 Multiple intrusion and deformation recorded by gabbros near the F915/ E907 intersection east of Chandria   This stop investigates two closely spaced localities along the F915 between Chandria to the west and Polystypos to the east, close to where it is intersected by the E907 coming south from Lagoudera. Park on the northern side of the F915, a short distance east of the intersection with the E907.

The view to the immediate north from the parking area, which is on gabbro, reveals that the sheeted dykes are striking towards the gabbro and are abruptly intruded by it. The gabbro adjacent to the sheeted dykes cannot be seen, but has been mapped as hornblende-bearing and pegmatitic with variable texture.

Walk southeastwards down hill to the start of the sharp bend to the right (36501118 E, 3867164 N). The roadcut exposure of interest begins here, goes around the bend and ends at the sign warning of falling rock.

This section records at least three phases of gabbro formation and the intrusion of poikilitic olivine gabbro–wehrlite. The most interesting features occur on the bend, but it will be necessary to work back and forth along the section to see the full picture.

The earliest gabbros are best examined at the start of the section. They are grey and they display a marked subhorizontal foliation, defined by the alignment of plagioclase and pyroxene crystals, which is the result of deformation and is not primary igneous layering. Close inspection will reveal that many of the thin discontinuous bands that define the foliation are in fact highly flattened fine-grain gabbro xenoliths (Fig. 2.14). These inclusions are completely recrystallized and may contain an even earlier foliation attributed to deformation. These two foliated gabbros have been intruded by dark brown-weathering olivine gabbro, which cuts across the exposure and contains abundant xenoliths, especially along its upper and lower margins. This latest gabbro is undeformed and it displays textural and compositional heterogeneity, as it contains some pegmatitic patches and may grade into, or be associated with, poikilitic plagioclase wehrlite. The latter lies at the top of the roadcut, but may be examined in fallen blocks by the side of the road, and may be part of a much larger body cropping out in the area.

Now walk back up to the parking area and westwards to the roadcut exposure marked by the large roadsign for the F915 to Chandria and Troodos (36 500807 E, 38 67215 N). The gabbros here are obviously layered and they vary from dark olivine-bearing varieties through to pale varieties.

Figure 2.14 Flattened pale fine-grain gabbro xenoliths within deformed medium-grain gabbro (stop 2.10). The coin is 28 mm across.

Layer width varies from 10 cm to more than 2 m, and mineral grading is obvious from the disappearance of olivine over a few centimetres upwards from the base of a layer. The gabbros do not appear to preserve a deformation foliation and may therefore represent a layered unit formed late and very high up in the plutonic sequence. Layers pinch and swell, some being discontinuous, and at least one layer displays isoclinal folds on a wavelength of about 10 cm. Much of this structure may have formed during compaction and slumping of the crystal mush within a pluton.

The layered gabbros are cut by a clinopyroxene-phyric doleritic dyke that appears to be chilled at both margins, although these show some evidence of shearing. The orientation of the dyke is the same as that of some minor reverse faults, and the dyke and these faults mark offsets of the layers. Taken together, the two road exposures portray the variability and complexity of the plutonic sequence at a very high level.

If time permits, it is well worth visiting the monastery of Panagia tou Araka ("Our Lady of the Pea") in its wonderful setting northwest of the village of Lagoudera, only a few kilometres north on the E907. It is a traditional mountain monastery with a domed roof hidden under a steep tiled cover that almost reaches the ground. Inside are splendid frescoes dating from 1192, which were restored between 1968 and 1973 to remove centuries of grime derived from candle smoke.

Stop 2.11 Diorite and plagiogranite north of the F915 between Alona and Fterikoudi This stop is located where the F915 passes over a prominent ridge situated about equidistant from Alona to the southwest and Fterikoudi to the east. At the summit, the road bends sharply and there is a parking area adjacent to a water pump and concrete water tank on the outside of the bend (36 504897 E, 38 66562 N). From here, walk northwards for several hundred metres along a gravel track in order to examine exposures of iron-stained quartz diorite and then white plagiogranite.

Diorite and plagiogranite represent the very last rocks to form during fractional crystallization of basaltic magma and, therefore, they usually occupy the tops of plutons. Because of the domed structure of the ophiolite and the resistance of these siliceous rocks to weathering, they characteristically crop out along prominent ridges close to the sheeted dykes. Views down slope reveal that gabbro lies around the diorite and plagiogranite, and the latter are believed to be part of a larger body of these rocks, which formed when fractionated magma intruded into the surrounding gabbro. The parking area lies on the intrusive contact, as gabbro is exposed on the inside of the bend opposite.

Along the first exposure there is a brown diabase dyke cutting iron-stained quartz diorite. The diorite comprises mainly quartz, feldspars and amphibole, but black crystals of magnetite are also abundant and they suggest that concentrations of iron and oxygen were high during the latest stages of magma crystallization. The presence of secondary green epidote in the diorite points to a high activity of water in the magma. This water reacted with primary minerals during the final stages of crystallization, causing epidote and, possibly, quartz to form. The water may have been contained within the magma, but more probably it was liberated during intrusion of magma into hydrated sheeted dykes. Farther along the track there is more diorite and then very pure white plagiogranite containing occasional green patches of epidote.

For those interested in inspecting altered plagiogranite, return to the parking area and walk a short distance down hill to the southwest towards Alona. Take the first dirt track on the right and follow this to the northwest to where it ends on a ridge occupied by plagiogranite comprising plagioclase, quartz and epidote (36 504741 E, 38 66592 N).

Stop 2.12 High-level gabbroic–dioritic intrusion into sheeted dykes north of the F915 between Fterikoudi and Askas   The stop lies along the F915 between Fterikoudi to the northwest and Askas to the south. Within 700 m southeast of the turning into Fterikoudi there are two wooden pylons on the northern side of the F915 close to a bend, and it is best to stop at the western pylon opposite a cutting into sheeted dykes (36 506859 E, 38 66145 N).

The cone-shape peak to the north across the valley comprises sheeted dykes with an attitude similar to those in the cutting along the road (i.e. steeply dipping and striking roughly north–south). However, Fterikoudi to the northwest sits on gabbro, so the gabbro/sheeted-dyke contact is sharp, which is a consequence of plutons intruding into the sheeted dykes. A clear example of this relationship may be seen in the middle of the cone-shape hill to the north, where there is a large, apparently funnel-shape, intrusion of gabbro and diorite into sheeted dykes (Fig. 2.15). The top and eastern contacts are obvious, but the western contact is more complex and it extends farther west than the view suggests. The top of the intrusion is composed of diorite and evolved gabbros, and these pass down, on the western side at least, into olivine gabbro that contains xenoliths of sheeted dykes. Patches of wehrlite also occur towards the lower part of the intrusion. The top of the intrusion shows a sharp contact with the overlying sheeted dykes. Xenoliths of sheeted dykes along this contact have been amphibolitized and recrystallized as a consequence of thermal metamorphism. Only rarely do dykes either cut the contact or arise directly from the intrusion to become part of the sheeted dykes.

Parts of the intrusion may be examined in more detail by heading about 250 m west towards Fterikoudi and then turning down hill towards the

Figure 2.15 Near-vertical sheeted dykes intruded by a funnel-shape body of gabbro and dior-ite that occupies the centre of the hillside (stop 2.12).

splendid little church of Agias Christinis. The narrow road cuts down through the sheeted dykes and into xenolith-bearing olivine gabbros that may be part of the intrusion. An eastward route can be picked across the hillside to the main body of the intrusion.

Chapter 3

The Troodos ophiolite: upper section

Introduction

The upper part of the crustal section of the Troodos ophiolite is made up of the sheeted-dyke and volcanic sequences and the Perapedhi Formation. These units surround rocks of the mantle and plutonic sequences exposed higher up in the Troodos massif (see Figs 2.2, 2.3). The upper part of the ophiolite preserves a complex history of magmatic activity, faulting, hydrothermal circulation and mineralization that occurred during seafloor spreading, and it is these processes that are investigated here.

The sheeted-dyke sequence

The sheeted dykes cover about half of the outcrop area of the entire ophiolite and are one of the most striking geological features on the island. At least several hundred metres of vertical section in the middle part of the unit is made up entirely of sheeted dykes that clearly demonstrate they were formed by repeated injections of younger dykes into older ones, to the complete exclusion of any other country rock (Fig. 3.1). The dykes strike broadly north–south and crop out for about 90 km across strike, showing 90 km of extension of the oceanic crust preserved in Cyprus. This is the most graphic demonstration of the reality of seafloor spreading to be seen anywhere in the world.

Put simply, the sheeted dykes represent the channels by which magma was fed from underlying chambers up to overlying volcanoes on the sea floor (see Figs 2.4, 2.5), but their upper and lower contacts are complex. The base of the sheeted dykes is usually intruded by plutons (stop 2.12), but dykes do intrude plutonic rocks in some places (stop 2.9), and in yet others the dyke and plutonic rocks are separated by a low-angle fault (stop 3.9). There are also very rare occurrences where gabbros can be seen

Figure 3.1 Steeply dipping sheeted dykes below the fire lookout on the eastern end of Madari ridge (stop 3.3). Note the cross jointing in the more prominent dykes.

directly feeding dykes (stop 2.12). The upper contact between the sheeted dykes and the overlying volcanic sequence is marked by a transitional unit of dykes intruding lavas, termed the Basal Group. This unit is made up of at least 50 per cent dykes, but usually more than 80 per cent dykes. Although a few screens of lava are found well down within the dykes and a few dykes penetrate far up into the lavas, the zone containing 20–80 per cent dykes is thin, usually no more than 200 m thick in a vertical section.

Rotation and faulting of the sheeted dykes and graben formation
Locally the dykes in the sheeted-dyke sequence are almost all nearly parallel to each other and have similar strike and dip, but on a more regional scale the strike and the dip change in a systematic way. Broadly speaking,

the strike of the dykes throughout the ophiolite is north–south, but there are large deviations from this direction. For example, the strike of the dykes north of the Arakapas fault zone swings around clockwise until it is nearly east–west and parallel to the fault. A similar 90° swing, but in the opposite direction, is seen in the extreme east of the ophiolite, close to the Mathiatis mine (stop 3.15). In other places there are swings of tens of degrees over a few kilometres. These changes in strike may be the result either of rotation of the stress field into which the dykes were intruded or of tectonic rotation of the dykes after intrusion (as is the case for dykes north of the Arakapas fault zone). Both of these processes might occur close to the ends of segments of a spreading zone.

The sheeted dykes are commonly tilted away from the vertical, so that over broad areas they dip either east or west at a low angle, in places close to horizontal. These shallow-dipping dykes must have been tilted after intrusion, but the amount of their rotation is usually far greater than can be seen in the lavas overlying them. This indicates that rotation must have taken place very early during the eruption of the lavas that cover the dykes. Tilted dykes of a consistent dip direction usually occur in elongate domains that run parallel to the local strike of the dykes and are up to several kilometres wide across strike. The boundaries between the domains may be faulted, but dykes of one orientation can be found intruding those of another close to a domain boundary. Where this is seen, the younger dykes are usually less tilted than the older ones.

There are three models that account for the rotational tilting of the domains of dykes (Fig. 3.2). The best known involves slippage of dykes along listric normal faults, but rotation driven by a domino-like behaviour or loading and subsidence is also possible. Whichever model is correct (and all three may be valid at different places in the ophiolite), the tilting

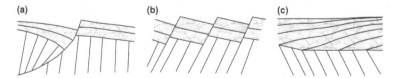

Figure 3.2 Three models for the rotational tilting of domains of sheeted dykes (in all cases the spreading axis is to the left). **(a)** Dykes rotate by slippage along arcuate listric normal faults that cut down steeply through the sheeted dykes and curve around to bottom out in major flat shear zones at depth, usually at the brittle–ductile transition that is the contact between the sheeted dyke and plutonic sequences (stop 3.9). If the fault throws down towards the spreading axis, rotated dykes dip towards the axis. **(b)** Dykes act like blocks of dominoes, bounded by steep faults, which rotate as dominoes do, by sliding past each other and sinking into softer, hotter rocks of the plutonic sequence below. As in (a), if the faults throw down towards the spreading axis, the dykes will dip towards the axis. **(c)** Dyke rotation driven both by loading from growth of the pile of overlying volcanic rocks and by deformation of underlying hot soft plutonic rocks. Unlike (a) and (b), dykes rotate to dip away from the spreading axis.

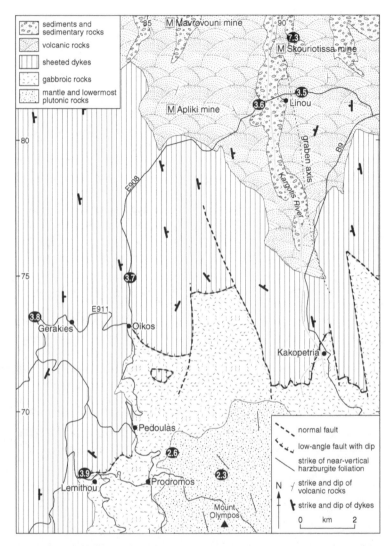

Figure 3.3 Simplified geological map of the Solea graben compiled from Wilson & Ingham (1959), Varga & Moores (1990) and Cyprus Geological Survey Department (1995). Faults at the sheeted-dyke/volcanic contact have been omitted for clarity. The numbers around the border are from the WGS84 UTM grid, where, for example, easting 85, northing 80 corresponds to the full grid reference 36 485 000 E, 38 80 000 N.

must have happened very close to the spreading axis, as tilted dykes are usually overlain by less tilted or untilted lavas.

Observations of the sense of dyke rotation have led to the identification of three graben-like structures on the northern side of the ophiolite, each of which is defined by two large dyke domains dipping towards one another either side of a central axis (see Fig. 2.2). The westernmost and largest of these is the Solea graben, which is centred along the valley of the Kargotis River north of Mount Olympos (Fig. 3.3). In the northern part of this structure, both the base of the lavas and the base of the sheeted dykes are downfaulted into a north–south rift about 3 km wide. The lavas thicken into this rift, so that the graben structure is scarcely present at the top of them. In the southern part of the graben there are outcrops of an extensive flat-lying detachment surface at, or close to, the contact between the sheeted dykes and the plutonic sequence. This is interpreted as the surface along which many of the listric faults bottom out. Still farther south, the steep foliation preserved in the mantle rocks around Mount Olympos may have been produced when mantle welled up beneath the graben. Taken together, these and other field relations suggest that the Solea graben may represent an extinct spreading centre, abandoned as the ridge jumped either to the east or to the west. However, extinct spreading centres on the sea floor generally lie within major valleys, a kilometre or more deep, within which plutonic rocks crop out, and such a valley is not seen in the Solea graben.

The Solea graben contains three of the largest copper mines in Cyprus, which collectively produced 75 per cent of the total copper and 40 per cent of the total mass of metal sulphides mined on the island. The large massive sulphide deposit of Skouriotissa lies just east of the graben axis, whereas the similar deposits of Apliki and Mavrovouni lie west of the axis (Fig. 3.3). These deposits and associated hydrothermal alteration permit conclusions to be drawn about the relative timing of igneous activity, hydrothermal circulation and faulting. The Skouriotissa deposit formed by black-smoker activity at the very top of the lavas, being covered only partly by a few tens of metres of lava, and must have formed late in the volcanic evolution of the graben. Conversely, the black smokers of Mavrovouni were covered by 100–200 m of lava, probably making them earlier than those at Skouriotissa, whereas those at Apliki lie near the base of the volcanic sequence and must, therefore, be earlier still. Thus, hydrothermal activity must have been going on throughout the accumulation of the lavas. Deeper down, hydrothermal alteration of the sheeted dykes has been shown mainly to pre-date the rotation of the dykes during faulting, but it must have continued to some degree during faulting and rotation, as some of the fault rocks are hydrothermally altered. This faulting did not rotate the upper lavas and, therefore, it must have been complete before they were erupted. Taken together, field relations in the Solea graben and elsewhere show clearly that, during construction of the Troodos crust, igneous activity, hydrothermal circulation and faulting very generally occurred in this order, but these processes also overlapped to a very large extent and there were cycles of these events.

Epidosites

In many outcrops of the sheeted dykes, one or two dykes will not be the normal bluish-grey or brown of most of the outcrop, but will be shades of yellowish or brownish green. This change of colour results because the dykes have been partly or wholly replaced by pale yellowish-green epidosite, a rock composed predominantly of epidote and quartz. In some places, broad regions of the sheeted dykes have been extensively altered in this way. The largest of these regions occurs within the Solea graben, where the shallowest epidosites are found around the village of Linou, due south of the Skouriotissa mine (Fig. 3.3), from where the region runs south for well over 10 km.

Alteration to epidosite involved major mineralogical and chemical changes in the original mafic rocks. Epidosites are extraordinarily enriched in calcium and depleted in magnesium and sodium; they and the rocks around them are also considerably depleted in copper, zinc and manganese, which are elements enriched in black smokers. The minerals formed during alteration contain abundant tiny inclusions of fluid that have about the same salinity as sea water and were trapped at about 350°C, the same temperature as black-smoker fluids. All of these observations suggest that epidosites mark the root zones of the hydrothermal systems that give rise to black smokers and sulphide deposits on the sea floor. Consequently, epidosite-rich regions represent the hydrothermal reaction zones where large volumes of cool sea water passed through sheeted dykes and became transformed into the hot metal- and sulphur-rich acidic fluid that eventually emerged from black-smoker hydrothermal vents.

Investigating the sheeted dykes

Representative parts of the main body of the sheeted-dyke sequence are examined in detail at stops 3.1 to 3.3. These stops are relatively close to stops 2.9 to 2.12 in the upper part of the plutonic sequence, and these can all be combined into a single transect. The Basal Group is the focus of stops 3.4 and 3.5, and epidosites are investigated at stop 3.3 and in the Solea graben at stops 3.7 and 3.8. Stops 3.5 to 3.9 may be combined into a comprehensive excursion to examine structural and hydrothermal aspects of the Solea graben (Fig. 3.3), which may take the best part of a day if studied in detail. If possible, combine this excursion with a visit to the Skouriotissa mine (stop 7.3; Fig. 3.3).

Stop 3.1 Sheeted dykes at the site of the Cyprus Crustal Study Project borehole CY4 at Palaichori The village of Palaichori is probably the best place to begin an examination of the sheeted dykes. It lies just off the E903 between Agros to the west and Apliki to the east. Coming from Agros on the E903, pass the left turnoff to Askas, Fterikoudi and Alona, and almost immediately afterwards turn left onto the E931 into Palaichori. From Apliki, the latter turning is on the right by a café, about 2 km on from the first entrance to Palaichori. Once on the E931, drive past the café on the right and park just beyond it opposite a roadcut (36 508423 E, 38 63764 N).

Walk onwards to the outside of the next bend, to where a pipe with a red padlocked cap sticks out of the ground. This is the top of the CY4 borehole, drilled by the Cyprus Crustal Study Project in the 1980s. It is 2263 m deep and it penetrated through 700 m of sheeted dykes before entering a section of complex plutonic rocks. The contact between these two units is marked by a sheared and altered zone that may represent a detachment surface.

Between the borehole and the café is a superb roadcut section of sheeted dykes. This section is an excellent place to begin detailed work on the dykes, because there are few faults and the dykes are nearly vertical and normal to the roadcut, they are little deformed, they show very good chilling relationships and they are only slightly weathered (so that the dykes are a bluish-grey colour in outcrop).

The most obvious features of the exposure are prominent long vertical joints and shorter joints at right angles to these (Fig. 3.4). The vertical joints may at first seem to be the margins of dykes, but this is not necessarily the case. Look more carefully at the outcrop, if possible with a piece of chalk to mark the dyke margins. Start by finding a dyke with two chilled margins. The chilled margin is an asymmetric contact, because on one side the rock has a coarse grain and does not show a change in grainsize towards the contact, whereas on the other side the rock has very fine grain indeed (chilled) and rapidly coarsens away from the contact. The fine-grain side is the margin of a younger dyke that has intruded into an older dyke. Mark the margin with chalk and draw an arrow pointing towards the centre of the younger dyke. Label the younger dyke A and the older one B. Now follow across dyke A and mark the other chilled margin in the same way as before. The piece of dyke beyond this chilled margin must also belong to dyke B. Continue in the same direction, labelling dykes as before, adding new letters as new dykes are identified. If the section reaches a fault, go back to the start point and work in the opposite direction. After 10–20 margins have been found, draw up the section, which should look something like Figure 3.5.

The section will show that most dykes have both chilled margins

Figure 3.4 Section of vertical sheeted dykes exhibiting four obvious chilled margins, indicated by arrowheads (stop 3.1). Close examination of sections like this will often reveal additional less obvious chilled margins. Chilled margins have finer grains and are usually darker in colour than the main bulk of the dyke interior. The hammer is 30 cm long.

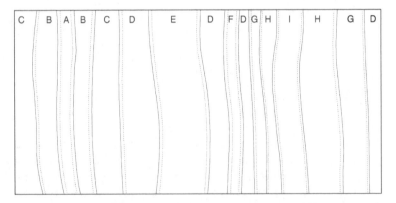

Figure 3.5 Schematic cross section of an exposure of sheeted dykes, probably about 10–20 m wide. Solid lines are the edges of dykes, with dotted lines showing the position of the inside edges of chilled margins. Dykes A, E, F and I have not been intruded by any other dyke, so each of these is complete and exhibits chilling on both margins. Dyke B has been intruded by dyke A, and dyke H by dyke I. Each part of dykes B and H, therefore, preserve only one chilled margin, as do the parts of dyke G, which has been intruded by dykes H and then I. The splitting of G by H, and of H by I, illustrates spreading in the sheeted dykes. Dyke D has been cut by many dykes and has been split into at least four parts in the section, and only one part has a chilled margin. Dyke C is the oldest and has no chilled margin in the section. In reality, some dykes and some parts of dykes are very thin indeed, because dykes often intruded along, or very close to, a pre-existing dyke margin.

present, even though a dyke may have been intruded by one or more other dykes, which may in turn have been intruded by still younger dykes. About one in ten dykes will have only one margin present in a section, with the other margin displaced outside the section, perhaps even to the other side of the spreading axis.

It is important to go to all of this trouble, because it is the only way to demonstrate that the sheeted sequence at this outcrop is in fact made up entirely of dykes. At first sight it appears that the coarse screens between some of the dyke margins are screens of plutonic rocks. By making a detailed section it is possible to show instead that the screens are slices of the centres of individual broad dykes, the chilled margins of which can be found farther along the outcrop.

Stand back on the opposite side of the road to take a broader view of the outcrop. It should now be possible to see the more prominent chilled margins as well as the faults that cut the outcrop. Note that the faults, which all appear to have a small displacement, are nearly parallel to the dyke margins, and often contain quartz veins. The quartz is in places associated with pyrite, suggesting that some hydrothermal fluid has risen along the faults, even though the intensity of hydrothermal alteration is not great. Those who have examined the entire roadcut may have found some bodies of plagiogranite or diorite.

Stop 3.2 Sheeted dykes along the E906 north of Platanistasa This stop lies along the E906 between Kato Moni and Platanistasa, about 9 km south of the road junction with the E907 to Agia Marina and 5 km north of the junction where the road to Platanistasa branches from the road to Polystypos. It is a 20 m-tall east–west roadcut, facing south, with dykes dipping steeply to the east (36 505822 E, 38 71448 N). This is an excellent outcrop of sheeted dykes that may be used as an alternative to stop 3.1, especially as it shows a wider range of features.

At the western end of the roadcut is a section of sheeted dykes that may be examined in the way described in stop 3.1 to identify chilled margins and hence show that the unit is composed entirely of dykes. Farther to the east it is possible to see some gently crosscutting relationships in the dykes, and some more radically crosscutting sheets. It is worth examining these to determine if there is any systematic relationship between the dip of the dykes and their relative ages.

Some dykes near the centre of the roadcut and towards its eastern end are green, showing that they have been replaced by epidosite and, thus, have had a greater degree of hydrothermal alteration than the other dykes around. Note that this more intense alteration is confined to individual dykes and has not affected neighbouring ones. Also at the eastern end is a major fault that cuts the section, and adjacent dykes are shattered. A fracture cleavage is developed in the fault zone, and secondary quartz veins with some pyrite cut the surrounding dykes. The fault runs roughly parallel to the dyke margins and does not seem to have a listric shape here.

Stop 3.3 Sheeted dykes and epidosites on the Madari Ridge north of Chandria
This stop ideally involves 2–3 hours of walking in the mountains to the north of Chandria (see Fig. 2.12), although it can be cut shorter if necessary. Not only are the sheeted dykes at their best here, but it is one of the great walks in Cyprus, with wonderful views, especially on clear days in the spring. The turnoff for the Madari Ridge lies at the summit on the F915 between Chandria (see stop 2.9) to the west and Polystypos to the east, where there is a crossroads with roads to Agros (south) and to a Cyprus Telecommunications Authority (CYTA) station (northwest). Take the road for the CYTA station, which at first runs across relatively flat terrain for about 1 km and then starts to climb at a hairpin bend. A large bus will have to stop here, about 1 km from the start point of the main walk. The flat terrain contains outcrops of the very top of the plutonic sequence, but the base of the dykes lies close to the bottom of the hairpin bend. At a distance of 2 km from the crossroads, the metalled road ends at the entrance to the CYTA station and there is a parking area with information boards (36 499912 E, 38 67643 N).

From the parking area, walk up the dirt road as it zig-zags through the sheeted dykes, in which there are sections both perpendicular and parallel to the dykes. In the sections perpendicular to the dykes, look for chilling relationships as described under stop 3.1. In some of the sections parallel to the dykes, there are joint surfaces that expose a single chilled margin over an area of a few square metres. These chilled margins are in many places wavy and grooved in a series of parallel undulations. The orientation of the axes of these undulations probably represents the direction of intrusion of magma in the dyke, as can also be seen from the imbrication of vesicles near dyke margins (see stop 3.10 for details). Note that the orientation of the undulations varies from dyke to dyke and is rarely vertical.

On the other side of the road are spectacular views of a bare hillside made up entirely of sheeted dykes (see Fig. 3.1), capped by a fire lookout perched on a ridge crest. To the east, about 2 km away, is another bare hillside of sheeted dykes, and beyond that are hills receding into the distance, all capped by sheeted dykes, demonstrating the vast extent of the sheeted-dyke sequence.

At the end of the dirt road is an information board with a map of a network of walks traversing the area. From here take the path labelled "Doxa Soi O Theos" that climbs to the crest of the ridge and leads along it. The path crosses excellent outcrops of sheeted dykes, with an occasional one altered to epidosite, and there are amazing views near and far: in the distance to the south is the salt lake southwest of Lemesos (stop 6.18); to the north the view stretches across the Mesaoria Plain to the Kyrenia Range; to the east is the knobbly backbone of sheeted dykes stretching all the way to the Kionia peak, topped with a television station; to the west are rocks of the mantle sequence forming Mount Olympos, scarred by the abandoned asbestos mine (see stop 7.18). Closer, below the ridge to the south, is rolling country dotted with villages on the plutonic sequence; the contact between the plutonics and the sheeted dykes runs along the base of the ridge. To the north is a narrow outcrop of plutonic rocks in the valley below and yet more hills of sheeted dykes before the cover of the volcanic sequence starts in the low ground near the Mesaoria Plain.

After 400 m on the path there is a sign stating "excellent viewpoint" (36 499149 E, 38 68068 N), beyond which there is abundant replacement of sheeted dykes by epidosite. For the critical field observations that can be made on epidosites, refer to stop 3.8. The good epidosite outcrops continue for another 400 m to a brown knob on the ridge just off the path (36 498747 E, 38 67912 N), where the transect ends. The trail continues after this and it is possible to make a 10–12 km round-trip back to the parking place using the marked paths, which is highly recommended if time, energy and drinking water permit. If not, return to the information board below the fire lookout and from here walk back down the dirt road to the lower parking place. An alternative and more atmospheric return follows a way-marked path around the base of the peak capped by the fire lookout and back to the parking place across spectacular crags of sheeted dykes.

Stop 3.4 The Basal Group along the E903 at Agia Koroni This stop is at a large roadcut, several hundred metres long, on the E903 between Arediou and Apliki. Coming south from Arediou, pass the turnoff to the west towards Agios Epifanios and drive on about 2 km to the first of the large roadcuts a short distance north of the small church of Agia Koroni. Here there is a good parking place on the east side of the road. From Apliki it is easiest to drive to the Agios Epifanios turnoff and then drive back to park, especially with a large coach.

The roadcut exposes sheeted dykes and pillow-lava screens of the Basal Group, the zone that represents the transition from the sheeted dykes to the volcanic sequence. At the southern end of the large roadcut (36 511930 E, 38 70818 N), the section is composed almost entirely of dykes that are highly rotated and dipping about 30° to the west. In places these dykes may be seen to be gently crosscutting, which is a common relationship seen in large outcrops. Only one dyke has a green epidotized core, whereas the other dykes are relatively unaltered. About 100 m to the north (36 511931 E, 38 70969 N), the proportion of dykes decreases to about 70 per cent and they are less rotated (dipping 55° west). In between these dykes is a large screen of pillows that should be studied to confirm that the section is the right way up. It is possible that the volcanic screens in the Basal Group were formed by molten lava flowing and falling into fissures in the sea floor, rather than by dykes intruding lava. Examine the contact between the dykes and the lava here to see if the relationship is intrusive or depositional. It is worth comparing these relationships with the clearly intrusive contacts preserved at stop 3.5.

Stop 3.5 Dykes and lavas at the axis of the Solea graben exposed along the E908 east of Linou   This stop lies along the E908, 2.8 km west of the junction with the B9 Lefkosia–Troodos road and about 300 m east of the first cross-roads for Linou and Katydata (see Fig. 3.3). Where the road cuts through a small hill, there is exposure and parking on both sides of the road (36 490788 E, 38 81362 N).

Before examining the roadcuts it is important to appreciate the struc-ture of the surrounding area to the east. From its intersection with the B9, the E908 runs westwards to Linou and cuts through a part of the volcanic sequence that has been downfaulted and tilted. Lavas and dykes exposed along this section of the E908 strike approximately north–south, and dykes dip west and lavas east. These orientations are attributed to tilting on the eastern side of the Solea graben and are quite the opposite to those found at stop 3.6 on the western side of the graben.

The roadcuts provide excellent sections through dyke-poor parts of the Basal Group, with screens of lava flows and pillows preserved between adjacent dykes (Fig. 3.6). Both lavas and dykes are evidently the right way up and they strike approximately north–south. The lavas dip very gently to the east, and the dykes dip steeply to the west, showing that these rocks have scarcely been rotated at all, which suggests that they lie at the graben axis (see Fig. 3.3). The dykes are clearly intrusive, as they have good chilled margins and they cut some pillows in half. Some dyke-in-dyke intrusion can be seen.

Near-vertical highly oxidized yellow and orange zones cut the lavas and many contain grey-weathering pyrite that is friable and sandy. These zones probably represent upflow pathways for hydrothermal fluids. They are cut by the dykes and this relationship shows that hydrothermal activ-ity started before magmatic activity had ended. The Skouriotissa mine (see stop 7.3) lies to the north directly along strike of the features in the road-cuts, suggesting that the altered zones in the cuttings may represent the fringe of the hydrothermal system that created the sulphides at the mine.

Figure 3.6 Near-vertical dark dykes intruded into and chilled against subhorizontal pillow lavas in the Basal Group (stop 3.5). The lavas have been hydrothermally altered and mineralized along near-vertical zones, which are also cut by the dykes. The notebook is 20 cm long.

Stop 3.6 Tilted lavas and dykes on the western flank of the Solea graben exposed along the E908 west of Linou This stop lies along the E908 a short distance up hill to the west of the Kargotis River, 1.8 km west of stop 3.5 (see Fig. 3.3). There is parking on the south side of the road by a gate, opposite a roadcut exposing lavas intruded by dykes (36 489100 E, 38 81332 N).

The lavas and dykes here are at a stratigraphical level similar to that at stop 3.5 and farther east along the E908, but they are obviously dipping in the opposite direction to their equivalents to the east. This relationship again suggests that stop 3.5 is located at the graben axis.

A more complete appreciation of the regional structure may be gained from the bridge over the Kargotis River, which runs from south to north (see Fig. 3.3). The present course of the river lies immediately to the west of the graben axis and the Skouriotissa mine. To the south rises the bulk of Mount Olympos, in which the mantle lithologies preserve a steep spreading-related foliation that strikes subparallel to the orientation of the graben axis. Between Mount Olympos and the bridge, there is a shallow-dipping detachment surface at the contact between the plutonic sequence and overlying sheeted dykes, a segment of which is examined at stop 3.9.

Stop 3.7 Listric normal faulting and epidotized sheeted dykes on the E908 north of Oikos This stop lies at the northern end of a very straight north–south stretch of the E908, 11 km from stop 3.6 and about 1.5 km north of Oikos (see Fig. 3.3). About 200 m south of where the E908 crosses the Setrachos River to follow the eastern side of the valley, there is a large cutting on the east side of the road, and a parking place on the west side immediately south of a house (36 484148 E, 38 74490 N). The exposure is best viewed in the afternoon.

From a distance it is clear that the whole cutting is made of variously grey and green sheeted dykes that are moderately to steeply dipping at the northern end and shallowly dipping or flat at the southern end. The cause of this arcuate structure is not obvious. It may simply be a product of folding, which is implied by dykes that appear to continue all the way around the arc. Alternatively, the arcuate apparent fold pattern of the dykes may not be a product of folding at all, but the result of movement of discrete slivers of sheeted dykes along one or more listric normal faults. Careful observation is required in order to determine the plausible mechanism.

If listric normal faulting is considered to be the mechanism, then it is possible that one of these faults is exposed at the base of the roadcut just north of telegraph pole KJ37 14 127. This fault dips down the road (to the north) and runs up the roadcut to the south. If this fault is listric and if the rotation of the dykes has taken place in the hanging wall of the fault, the steeper inclination of the dykes in the northern (hanging-wall) block would require that the steeper dykes have been rotated through horizontal and are now inverted with respect to their original attitude. The shallow inclination of the dykes in the south (footwall) would then originate from rotation in the hanging wall of an earlier listric fault that lies below the present level of the exposure. Such large rotations are possible in stacks of multiple listric faults. It is not possible to say if the roadcut exposes a major listric fault or splays off a larger fault located elsewhere. Nevertheless, exposures like this have been used as evidence to suggest that blocks of the upper crust slid towards the graben axis by rotating along several listric normal faults.

The green colour of many of the dykes is caused by the presence of epidote. The greenest dykes are composed of about half epidote and half quartz, with only minor amounts of other minerals. The dykes at the northern end of the outcrop seem to contain a higher proportion of epidote than those at the southern end. A late dyke cutting earlier faulted dykes may be seen along the base of the roadcut, just south of the fault (look for the cored dykes). This dyke is about 20 cm wide and has not been epidotized, although it cuts a dyke that has been replaced by epidosite. In general, within outcrops of dykes that include epidosites, it is possible to find dykes with more epidote cutting others with less epidote, and vice versa, showing that there was not a single period of epidosite formation affecting a large volume of already-mature dykes. Epidosites are examined in detail at stop 3.8 (to the west), but they also occur to the north, very close to two massive sulphide deposits that were exploited at the mines of Mavrovouni (the largest known orebody in Cyprus: 14 million tonnes of ore) and Apliki. Clearly, epidosite is a relatively common rock type in the western part of the Solea graben and it shows the large amount of hydrothermal circulation that this area must once have experienced.

Stop 3.8 Sheeted dykes and epidosites along the E911 west of Gerakies   This stop lies along the E911, about 2 km west of Gerakies and 2 km north of the intersection with the E912 (see Fig. 3.3). There is a large parking place on a sharp bend where there are wide views all around to the west and the north (36 480805 E, 38 73366 N). This is an excellent place to see epidosites, and hydrothermal enthusiasts should certainly not miss it.

From the parking area, walk south along the road for 200 m to a junction at a corner where a forest road to Atratsa goes down hill to the west (36 480822 E, 38 73149 N). The section beyond this corner contains intensely epidotized sheeted dykes dipping at 45° to the east (Fig. 3.7). Here it is possible to make critical field observations that bear on the genesis of the alteration. First, ensure that the section is made of sheeted dykes by identifying chilled margins in the way described at stop 3.1. This will show that some

Figure 3.7 Outcrop of epidosite within moderately dipping sheeted dykes (stop 3.8). In the centre of the picture is a dyke that is about 2 m wide, which has a highly altered core and stripes of darker and lighter alteration parallel to its margins. A 40 cm-wide dyke just above the striped dyke has a highly altered centre and wide dark (and hence less epidotized) margins. The notebook is 20 cm long.

of the dykes are wide and others are narrow, and that some dykes are subtly crosscutting. Next, examine the distribution of the epidote, given that, the greener a rock is, the more epidote it contains. Although there is some epidote filling joints and small faults, most of the epidote replaces the interior of individual dykes. Margins of dykes typically contain less epidote than dyke centres. In the wider dykes, stripes of epidote-rich and epidote-poor rock run parallel to the dyke margins, whereas the centres are more uniformly replaced (Fig. 3.7). Some dykes are grey and have very little epidote in them; others are bright green and have been entirely replaced by epidote and quartz. Note that the columnar cross joints that are so prominent in unepidotized dykes elsewhere are absent here. Where dykes cut across one another, the stripes of epidote-rich rock faithfully follow the curvature of the dyke margins. It is worth trying to decide whether, in general, more-epidotized dykes cut less epidotized ones, or whether the reverse is true.

Putting all of these observations together, the fluids that altered the dykes to epidosite must have penetrated the interior of the dykes and were not channelled along dyke margins, faults or joints. Because of the high permeability of cracks in rocks, it is thus likely that much of the fluid circulated before any faults or joints formed in the dykes. The systematic parallelism of epidote-rich bands to dyke margins shows that the fluids flowed at a time when the dyke margins and interiors were still distinct, and individual dykes could channel the flow. The extent of replacement shows that dykes must have experienced a large throughput of hydro-thermal fluid.

There is no systematic relationship between degree of epidotization and relative age of the dykes. A thoroughly epidotized dyke can be found intruding one much less altered, and vice versa, again indicating that the fluids were channelled along individual dykes. All of this field evidence suggests that the alteration happened very soon after the intrusion of each dyke, probably at a time when the centres of the dykes were still porous immediately after solidification and before any secondary minerals had filled these primary pores.

It is worth spending some time walking farther south along the road in order to confirm these relationships and to observe that the degree of epi-dotization generally decreases southwards. On the way back to the park-ing area, the road runs along a strike section of the epidotized dykes in which there is a small fault. A few hundred metres north beyond the park-ing area is a major fault zone that is highly epidotized. This zone may have acted as a channel through which hydrothermal fluids were focused after reacting with the dykes.

If time permits, it is worth travelling south to join the E912 and then west to the monastery of Panagia tou Kykkou and its impressive Byzan-tine museum (36 476440 E, 38 71260 N). The former is the richest and most famous of religious institutions in Cyprus, and one of the most celebrated in the Greek Orthodox world. The founder of the monastery was a hermit called Isaiah and he lived in a nearby cave in the twelfth century. Repeated fires have left nothing from the early history and everything now dates from 1831 or later. The monastery is closely linked to the Cypriot nation-alist struggle, and Archbishop Makarios III hid out here during the days he was involved with EOKA (the National Organisation for the Cypriot Strug-gle). His tomb may be visited on nearby Throni Hill, the very top of which affords fantastic views on a clear day.

Stop 3.9 A detachment fault between sheeted dykes and gabbros at Lemithou
With a large coach it is necessary to approach Lemithou from the west (see Fig. 3.3). This must be done by heading southwards on the F810 towards Treis Elies and Kaminaria from the E912 Pedoulas–Kykkou road between Pedoulas (to the southeast) and the E911 intersection (to the northwest). Follow the F810 over a col and down to a T-junction, noting excellent exposures of highly rotated and epidotized sheeted dykes, and turn left towards Lemithou at the junction. There is no convenient parking in Lemithou, so park just below (west of) the first houses on the outskirts of the village, in an open space (36 482436 E, 38 67553 N). The detachment

fault lies directly up hill from here, but it is best to approach it through the village. The parking area may also be reached by small vehicles approaching from Prodromos (to the east) and Palaiomylos (to the south), by following the signs to Lemithou and passing through the village to its western side.

From the parking area, walk up into the village, on the way examining excellent exposures of gabbro with pegmatitic patches, black hornblende veins and net veins of plagiogranite. About 10 m beyond the village taverna, take a concrete track on the left heading up hill; alternatively, walk farther up to a track about 100 m beyond the post office. Follow either track until the outcrop seen from the parking area comes into view. From a distance it can be seen that shallow-dipping sheeted dykes overlie gabbro (Fig. 3.8). At the outcrop (36 482550 E, 38 67644 N) it is clear that the dykes are highly epidotized, that the gabbro exhibits a highly varied texture and is strongly fractured, and that the contact between them is faulted. Be careful when examining the outcrop, as it is very unstable.

There is a strong contrast between the style of hydrothermal alteration above and below the fault, suggesting that the dykes were already epidotized before they reached their current position. Low-angle faults are often considered to be located at the brittle–ductile transition within the crust, where the fault rock and the footwall below it exhibit textures indicative

Figure 3.8 Outcrop of the Lemithou detachment (stop 3.9). Above the fault (dashed line) the hanging wall is made of sheeted dykes that have been epidotized and rotated nearly to horizontal. The underlying footwall consists of highly fractured gabbro. Note the lens of intensely fractured gabbro that has been transferred from the footwall to the hanging wall and is surrounded by strands of the fault. The prominent thick dyke is almost 2 m wide.

of intense granulation and shearing. At Lemithou both the fault rock and the footwall are predominantly deformed by brittle fracture, although there are signs of ductile fabrics in some of the clasts in the fault rock. Much of the brittle fracturing is associated with veins of the chalky white zeolite, laumontite, indicating temperatures of about 200° C.

The fault cropping out at Lemithou has been interpreted as a detachment surface along which arcuate listric faults flatten and sole out on the western side of the Solea graben (see Fig. 3.3). The fault would have thrown down towards the graben, so that the hanging-wall block of dykes would have slid east over the gabbro. Try making structural measurements to test whether a simple rotation about a north–south near-horizontal axis, coupled with sliding along the fault plane, could have brought the dykes from vertical to their present position, or whether a more complex structural history is required. Examine the orientation of slickensides in the main fault zone and the associated minor fault strands to see if they are consistent with a particular mechanism. Although the early fabrics are consistent with the fault having originally been located at the brittle–ductile transition within the crust, corresponding to the boundary between the sheeted-dyke and plutonic sequences, much of the movement on the fault must have taken place after the plutonic rocks cooled. Evidence from elsewhere shows that most of the fault slip must have occurred during eruption of the lavas, which is intriguing, as this implies that plutons cooled very rapidly during the evolution of the graben.

The volcanic sequence

The main outcrop of the Troodos ophiolite is surrounded by a unit of submarine volcanic rocks intruded by subordinate dykes that make up the volcanic sequence (see Fig. 2.2). This sequence is best developed along the northern margin of the ophiolite, which is also where good examples of umber and deep-marine sediments of the Perapedhi Formation are found filling in depressions on the lava surface. Distant views across the very top of the brown- and grey-weathering volcanic sequence clearly reveal the small amount of deformation that this part of the ophiolite has experienced since it formed 90 million years ago (Fig. 3.9).

Traditionally, the volcanic sequence has been subdivided into the Basal Group, the Lower Pillow Lavas and the Upper Pillow Lavas. The Basal Group is the transitional unit between the sheeted dykes and overlying volcanic rocks, examined in the previous section specifically at stops 3.4 and 3.5. The middle and upper parts of the volcanic sequence are made up of the Lower and Upper Pillow Lavas, but these terms are not used here

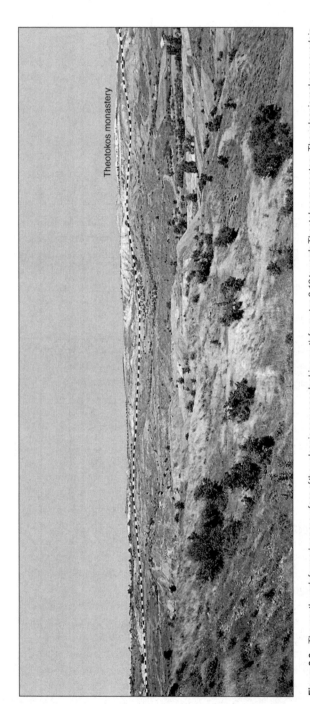

Theotokos monastery

Figure 3.9   The mostly undeformed upper surface of the volcanic sequence looking north from stop 3.18 towards Theotokos monastery. The volcanic rocks are overlain (contact marked) in the distance by gently north-dipping white chalks of the Lefkara Formation.

because much of the sequence is not made of pillow lavas and it is much more structurally and chemically complex than these two names imply. In fact, throughout the volcanic sequence there are four main types of rock that can be seen in the field:

- Pillow lavas (Fig. 3.10a). These lavas have the shape of bulbous pillows, usually ranging in size from 20 cm to over 1 m across, having flat or domed tops, and bases that are moulded over the tops of underlying pillows. Many pillows were rapidly quenched when they were erupted and, consequently, they have glassy chilled margins or their altered equivalents. Cooling joints may be developed and these radiate out from the centres of pillows.
- Sheetflows (Fig. 3.10b). These are laterally extensive continuous flows, each with a rubbly top and base, and a massive centre that is usually columnar jointed. They should not be confused with sills, which have chilled upper and lower margins and may subtly crosscut the lava stratigraphy. Sheetflows usually contain the conspicuous bluish-green alteration product celadonite, which is a type of mica. They are believed to have formed at eruption rates that were higher than those that produced pillow flows.
- Hyaloclastites (Fig. 3.10c). These are volcanic sediments made up of sand-size fragments of basaltic glass, which probably represent deposits formed by fire-fountain activity on the sea floor. These sediments are usually associated with sheetflows and are intruded by tongues of basalt in many places.
- Pillow breccias (Fig. 3.10d). These are piles of fragments of pillows that may have formed either as primary deposits at the fronts of some pillow flows, or as secondary deposits of seafloor talus at the foot of fault scarps or steep submarine slopes. The primary or secondary origin is often difficult to determine.

The appearances and relationships of the four main types of volcanic rocks in the field are both complex and evocative. Understanding their origin and mode of emplacement has been greatly aided by direct observations of lava flows at modern seafloor-spreading centres. Most eruptions occur from a narrow fissure zone, but some are erupted from off-axis vents up to a few kilometres from the spreading axis. Although some flows reach as far as 1–2 km or more from the eruption site, many lavas flow for only a few 100 m. Flow thickness and morphology vary according to eruption rate and temperature, distance from the eruptive centre, and topography of the sea floor, which may be highly irregular. Accordingly, individual flows may be sheet-like in some places and pillowed in others.

Further complexity throughout the volcanic sequence is provided by the occurrence of dykes and sills. From studies of modern spreading

Figure 3.10 The four main types of volcanic rocks found in the Troodos ophiolite. **(a)** Pillow lavas along the Akaki River between stops 3.11 and 3.12. Pillows, each one about 1 m across, are outlined by their dark glassy margins, now replaced by low-temperature alteration minerals. These pillows are near horizontal and the right way up. **(b)** Thin sheetflows exhibiting massive columnar-jointed centres and rubbly tops and bases (stop 3.11). Preferential erosion of the tops and bases makes it difficult to define the margins of each flow. The flow units in this image may all be lobes of one much larger lava flow. The horizontal field of view is about 10 m. **(c, opposite)** Hyaloclastite intruded by tongues of basalt (some of them labelled B) near stop 3.10. The hammer is 30 cm long. **(d, opposite)** Pillow breccia from close to the top of the volcanic sequence (stop 3.12). Some of the pillow fragments are small and angular; others are almost complete, although rather fractured (P). Chilled pillow margins are altered to pale clay minerals. The hammer is 30 cm long.

centres, it has been shown that some dykes feed eruptions directly above a magma chamber, whereas others intrude for up to tens of kilometres along axis from a chamber. In the ophiolite, dykes are quite rare in the volcanic pile, and their abundance decreases very rapidly above the top of the

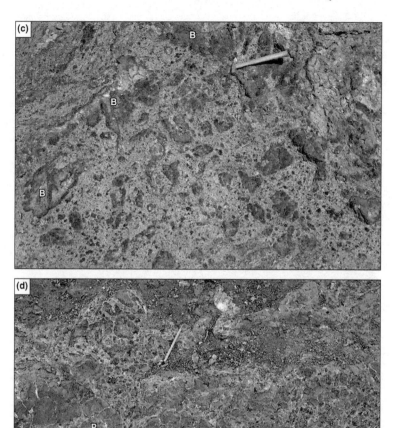

sheeted dykes. This rapid disappearance and rarity of dykes may at first seem surprising, because almost all of the volcanics must have been fed by dykes tapping magma chambers in the plutonic sequence. An abrupt lava–dyke transition results because the zone of dyke intrusion at the spreading centre is far narrower than the extent of lavas flowing farthest from it. Such a situation has been recorded at the East Pacific Rise, where dykes are intruded into a 100–200 m-wide zone of fissuring, but the lavas they feed flow away from this zone for more than 1 km. Consequently, dykes in the zone of fissuring become covered by the short flows and the proximal parts of longer flows they feed. As these dykes and flows move away from the spreading centre, they are covered successively by progressively more distal parts of extensive younger flows erupted from the same centre.

The lavas are mostly aphyric, but others contain phenocrysts of olivine, clinopyroxene and chromite. In some pillows, olivine phenocrysts are very abundant and make up more than 50 per cent of the rock and, in pillows and sheetflows, olivine may have settled to the base of the lava. The lavas commonly contain few vesicles, probably because gases were held in solution in the magma by the pressure of the overlying sea water. However, some sheetflows do contain abundant vesicles, apparently because they trapped steam as they spread over and heated the sea floor.

Many original igneous features of the volcanic sequence have been overprinted by seafloor weathering. Typically, at the top of the volcanic sequence, cold oxygen-rich sea water percolated through the lavas over millions of years and altered them. These rocks are coloured orange by iron oxides, and contain brown oxide-rich material replacing olivine phenocrysts, and the alteration product palagonite in place of quenched basaltic glass. The lavas are also cut by abundant fissures that are now filled with calcite and other low-temperature minerals. Less water seems to have penetrated deeper into the lava pile, and temperatures were probably a little higher there than at the surface. Some dark fresh basaltic glass remains, and alteration minerals include celadonite (once exported from Cyprus as a pigment) and silica in the form of agate.

There are many possible stops in the volcanic sequence, five of which are presented here. All are excellent and highly recommended for different reasons, but, if time is short, then stops 3.10 and 3.13 are the ones to choose. Stops 3.4 and 3.10 to 3.12 may be combined to provide a complete and convenient section through the volcanic sequence centred along the E903 between Apliki and Arediou. Other places where good exposures of Troodos volcanic rocks may be examined are stops 3.5, 3.15, 3.17, 3.18, 4.1, 6.2, 7.3, 7.4 and 7.12.

Stop 3.10 Volcanic rocks intruded by dykes along the Akaki River southwest of Klirou This stop is one of the truly classic geological sites in Cyprus. From the E903 between Agia Koroni (stop 3.4) about 5 km to the south and the E903/E905 intersection about 4 km to the northeast, turn eastwards onto the F962 towards Klirou and park just beyond the bridge over the Akaki River (36514144 E, 3874322 N). From here, follow the dirt road off to the right for a short distance before taking a well trodden path on the right that leads up to a high point on the southern side of the river.

The view north across the river to the canyon wall is spectacular and

Figure 3.11 Swarms of dykes intruded into pillow lavas and hyaloclastites in the lower part of the volcanic sequence close to the Basal Group (stop 3.10). The largest pillows are about 1 m across.

shows a section through the lower part of the volcanic sequence (Fig. 3.11). A lower unit of hyaloclastites and an upper unit of pillow lavas are cut by swarms of near-vertical dykes. These dyke swarms make up perhaps 30 per cent of the outcrop and are themselves cut by inclined dykes. Each swarm consists of several dykes, each intruded close to the same axis. This site is near to the level at which dykes become abruptly abundant on passing downwards into the lowest part of the volcanic sequence.

It is worth making an annotated sketch of the exposure, including an indication of the way-up, before heading down via the bridge to examine the rocks in the canyon more closely. At the first swarm, verify that there are indeed several dykes here by looking for chilled margins between them. Some of the dykes show beautiful columnar cross jointing. One of the dykes shows excellent evidence for the direction of flow of magma, as it contains abundant elongate vesicles close to its margin. By examining the dyke margin sideways on, it is possible to see that the elongation of the vesicles is nearly horizontal, which shows that magma was intruded from either the southeast or the northwest. This is interesting because, as the whole section is the right way up, as can be deduced from the structure of

79

the pillow lavas, it demonstrates that magma flowed laterally rather than vertically. It is easy to get into the habit of assuming that magma in a dyke always flows upwards, but it does not. Now look vertically down on the dyke margin to see that the vesicles overlap like tiles on a roof and do not run exactly parallel to the dyke margin. This was caused by the flow of magma in the dyke being faster in the centre than at the margins. From the sense of overlap it can be determined that the magma flowed from the northwest.

Climb over the dykes and look at the hyaloclastites beneath the pillow lavas. The hyaloclastite itself is the brownish material that is a somewhat altered pile of sandy basaltic glass. It may have been produced by submarine fire-fountaining, indicating that a seafloor vent may have been very close at the time of eruption. The glassy volcanic sediment has been intruded by tongues of basalt of a wide range of shapes. This basalt must have been injected into the hyaloclastite while it was still unconsolidated, because it has the same texture as is produced when basalt is intruded into wet sediment. The basalt has been partly altered to bluish-green celadonite, and vesicles have been filled with agate.

About 30 m farther southwest along the section, and about 10 m above the river bed, is a thin inclined dyke of basalt cutting across the hyaloclastite and the basalt bodies intruding it. This dyke has a thick margin of glass where it intrudes the hyaloclastite and a thin one where it intrudes the basalt. It has been claimed that this variation in margin width is a result of the dyke fusing water-saturated hyaloclastite during intrusion, but not melting the basalt. It is possible to think of other explanations: for example, the thicker glassy margins may record more extensive chilling of the dyke adjacent to the hyaloclastite.

A short distance still farther southwest along the river bed are outcrops of pillow lavas, some of which exhibit superb bifurcating lobes that enable the direction of flow to be determined. The lobes point in the general direction of movement of the lava front.

Stop 3.11 Lavas, dykes and faults along the Kamara River west of Malounta
Along the E903, 2–3 km southwest of Arediou and about 4 km northeast of stop 3.10, is a crossroads with the E905 heading west to Agrokipia and Mitsero and a minor road to the east to Malounta and Klirou. Follow this minor road into the valley of the Akaki River and park just east of the bridge (36516189 E, 3877305 N). The Kamara River is in fact the small stream coming through the canyon to the east, and it meets the Akaki River about 20 m north of the road bridge.

On the northern side of the canyon, just up from where it meets the road, is a slope of pillow lavas with good examples of black glassy margins.

At the entrance to the canyon there are a few thin sheetflows with well developed columnar centres and less obvious rubbly tops and bases (see Fig. 3.10b). Note the bluish-green veins and patches of celadonite in the flows.

Walk on the path on the southern side of the canyon for about 30 m and look across to the northern side (a better view may be obtained by climbing up the southern slope). The bottom part of the steep cliff is made of the same unit of sheetflows seen at the canyon entrance, but above it is a very different unit dominated by circular features up to 10 m across that exhibit radial columnar jointing. These resemble enormous pillows (sometimes called megapillows) and they may represent large lava tubes through which the higher flow was fed.

Continue walking eastwards and then around the corner to the right. Here the path and river bed lie on the hexagonal tops of sheetflow columns. Note the intriguing concentric bluish-green rings of celadonite within the columns. Farther on, the canyon bends left and there is an outcrop of hyaloclastite on the inside corner. This still contains a fair amount of fresh basaltic glass, but the brownish colour reflects the presence of alteration products such as palagonite.

Continue up stream and along a faint path above the river bed on the south side of the canyon. On the north side of the canyon is one of the classic outcrops of volcanics in Cyprus (Fig. 3.12; 36 516328 E, 38 77226 N). The majority of volcanics are sheetflows with rubbly margins and columnar

Figure 3.12 Fault (white line) cutting sheetflows and hyaloclastites (stop 3.11). The thick sheetflow to the left of the fault is apparently the same flow as one of the thin flows to the right of the fault, which demonstrates that faulting was occurring at the same time as lavas were erupting. A late low-angle intrusive sheet clearly cuts across the outcrop and probably the fault. The horizontal field of view is about 15 m.

81

cores. In the middle of the outcrop is a fault that has displaced the lower flows and has thrown down to the west. A thick flow on the western side has ponded against the fault scarp and then has overflowed it to the east. Alternatively, the flow may have come from the east and poured down the fault plane into the depression created by faulting, and then continued flowing west. Later flows cross the line of the fault with little if any displacement. A steeply east-dipping dyke lies just to the east of the fault and may have intruded along another fault. The whole outcrop is cut by an inclined thin sheet that is a dyke in places and a sill in others, and it is truncated by a dyke at its far eastern extent. The whole outcrop illustrates clearly the overlapping in time of active faulting and volcanism at a spreading centre.

On return to the Akaki River, you may walk northwards along the river, under the E903 road bridge, to stop 3.12. In the valley section the lavas closer to the Kamara River (altered deeper in the lava pile – perhaps 200–300 m below the sea floor) are greenish grey in colour and preserve fragments of fresh basaltic glass in pillow margins, whereas those farther north are the orange colour typical of lavas altered at or close to the sea floor (stop 3.12).

Stop 3.12 Pillow lavas, lava tubes and pillow breccia along the Akaki River west of Arediou This stop lies north along the Akaki River from stop 3.11, from where it can be reached on foot along the river valley. By road, from the E903 about 2 km southwest of Arediou and 1 km northeast of the junction with the E905 to Agrokipia and Mitsero (see stop 3.11), turn north onto the road to Agios Ioannis. After about 200 m, park just beyond the first bend in the road (36 516604 E, 38 78364 N) and follow a winding track westwards down to the river. About 20 m before the river, there are cliffs of Lefkara chalk on the right; these are the lowest sedimentary rocks overlying the lavas here and the contact is unconformable. The chalk is Oligocene in age (about 30 million years old), implying that the lavas were exposed at the sea floor for about 60 million years after eruption, presumably because they formed a local high on the sea floor.

Turn south along a faint track through olive trees. The flat area just here (36 516394 E, 38 78368 N) is the site of the Cyprus Crustal Study Project borehole CY1, which was drilled in the 1980s and penetrated several hundred metres into the lava pile. Some 50 m south of this point, climb up to the left onto a faint track to the lava crag above, which rises above an old water course, which once fed a nearby mill. The crag is made up of pillows coloured pale orange by seafloor weathering. The once-glassy margins of the pillows have been replaced by deep orange palagonite, and the shape of the pillows is outlined by these orange rims. The pillows contain phenocrysts of pale-green olivine (partly altered to iron oxides), emerald-green clinopyroxene and tiny black octahedra of chromite. The pillows cracked on the sea floor and the spaces were subsequently filled with calcite.

Now follow the water course to the right, and south around a corner, for about 30 m to a concrete water cover. Near here the lavas are still pillowed and many contain abundant olivine phenocrysts up to 10 mm across. In some of the pillows the olivine phenocrysts have accumulated at the bases of pillows, clearly having settled in the magma after the pillows had formed. In others, the olivine phenocrysts are concentrated at the centres of pillows, suggesting that these are not simple dead-end pillows but elongate lava tubes that probably fed other pillows down slope. Indeed, the elongate tube-like form is clear to see in some parts of the outcrop. The phenocrysts may be concentrated in the centre because a slug of crystal-rich magma was intruded into the tube when it had partly solidified.

Over the next 30 m along the water course, the pillows gradually break up into a pillow breccia, in which the angular sectors of individual pillows are recognizable as clasts in the breccia (see Fig. 3.10d). This gradual transition indicates that the breccia formed during an eruption, so that it is a primary volcanic product rather than part of a talus pile. After another 50 m along the water course, intact orange pillows are encountered again. From here it is possible to continue walking south to stop 3.11, effectively going deeper into the lava pile, as suggested by the change in seafloor weathering and alteration of the lavas.

Stop 3.13 Volcanic rocks and faulting southwest of Kato Moni About 1 km southwest of Kato Moni, one of the very best sections of volcanics is exposed at the junction of the E906 Peristerona–Platanistasa road with the E907 to Agia Marina. Park on the open ground opposite the road junction and start at the low cutting on the E906 (36 507743 E, 38 79194 N). The whole section, both here and along the E907 towards Agia Marina, contains a wide range of lava types adjacent to one another in excellent exposure and a fine example of a major fault cutting them.

The low roadcut exposes gently westerly dipping pillow lavas that exhibit classic flattened and rounded tops, and bases moulded into underlying pillows. After examining these, walk around the corner to examine the section along the road to Agia Marina (take good care, as traffic travels very fast along this road). The north side of the roadcut is usually the best to work on because of the better lighting. About 20 m from the road junction, there is a sheetflow extruded on top of pillows and covered by later pillows. Note that the base of the sheetflow is moulded over the tops of the underlying pillows, indicating that it is indeed a flow and not a sill.

Farther on, the pillowed unit becomes gradually fractured and red-dened in a wide damage zone associated with a major normal fault that dips to the east and throws down in the same direction, as can be seen from the fabric in the fault zone itself. This fault, which strikes parallel to the local dyke strike of north-northwest–south-southeast, must have formed at the spreading centre as one of the major faults active during the con-struction of the crust. Beyond the fault zone is an entirely different unit of lavas, which consists of an association of hyaloclastites and sheetflows. The sheetflows have excellent columnar-jointed centres and rubbly tops and bases. This unit is cut by a sill that dips gently to the east. At first sight this sill is very similar to the sheetflows because of its columnar-jointed centre, but it is distinguishable from them by its upper and lower chilled margins, and by its gently crosscutting relationship with the flows. The best hyaloclastites are seen at the far (west) end of the roadcut.

Stop 3.14 Olivine-phyric lavas and half-graben at Margi   Margi is a small village on the E901, 6 km south of Tseri and 3 km northwest of Kotsiatis. At the main bend in the E901 on the outskirts of the village, turn west into Margi, drive up a low hill, and turn left to follow a metalled road to the first turning on the right onto a dirt road. Park here (36 529382 E, 38 75133 N) in order to examine the geology south and west of Margi (Fig. 3.13).

Start by heading about 100 m northwest to the prominent dark knoll (K in Fig. 3.13), which is surrounded by lava flows, most of which are thin sheetflows that strike northwest–southeast and dip to the northeast. The knoll itself is columnar jointed, the columns dipping steeply to the south-west, and it is made of ultramafic lava that grades upwards to the north-east into an aphyric basalt. The ultramafic lava is packed with olivine phenocrysts, which form about 60 per cent of the rock. In the dark sand derived from the knoll, it is possible to find millimetre-size fragments of fresh olivine.

A clue to understanding the structure of the knoll comes from a point about 100 m to its south, where a sheetflow, about 5 m thick, lies above pillow lava. The sheetflow is dark and rich in olivine at its base, and it grades upwards through a thin (about 20 cm-thick) transition zone into an aphyric upper part. Following the flow to the southeast will reveal that the olivine-rich lower zone pinches out and the entire flow becomes aphyric. This is clear evidence that the basal zone formed by accumulation of olivine phenocrysts in a depression at the bottom of an otherwise aphyric lava flow. The same relationship can be proved many times over in other flows near by. The same gradational transition can be shown in the dark knoll already examined, but, if it formed in the same way, the knoll must

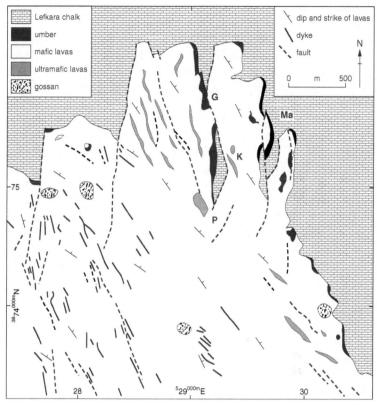

Figure 3.13 Geological map of the area around the village of Margi (stop 3.14), simplified from the unpublished maps by Richard Smith and Rex Taylor. Also marked are the locations of Margi (Ma), a dark knoll of ultramafic lava (K), an exposure of ultramafic pillow lavas (P) and the best exposure of a half-graben (G). Border numbers are from the WGS84 UTM grid.

represent a relatively deep depression in the base of a flow for such a large accumulation of olivine to have occurred. The knoll has also been interpreted as a plug intruded into pre-existing lavas and, therefore, careful examination of the field relations is critical for deciding between these modes of origin.

Now walk about 50 m up hill to the contact between the lavas and the overlying Lefkara chalk. The contact is in fact a fault scarp and the normal fault strikes north–south and threw down to the west to create a half-graben. This depression subsequently filled with umber and siliceous mudstone, and was then covered by chalk. The Margi area is cut by several of these half-graben, which are very similar to those seen at Theotokos monastery (stop 3.16), but excavations and the construction of a chicken farm have made these hard to see at Margi. More of the structure is visible if the contact between chalk and lava is followed up hill southwards, in order to gain a view northwards over the top of the chicken farm. A dip surface of lava can be seen to be covered by umber, and the fault that has been followed, forming the eastern side of the half-graben, continues under the chicken sheds. If there is time, it is worth walking north for about 500 m past the farm, where the half-graben and its fill of umber and chalk are seen more clearly (Fig. 3.13: G).

These half-graben can be shown to be later than the primary construction of the crust in this area. Figure 3.13 shows clearly that the strike of the lavas and the dykes is northwest–southeast, which seems to indicate the orientation of the spreading axis at the time of crustal construction in this area. A set of faults striking northwest–southeast can be mapped as being associated with this early orientation (Fig. 3.13). The half-graben faults run north–south and are clearly superimposed on the earlier structures, perhaps reflecting a minor adjustment of the orientation of the spreading axis.

About 250 m south of the viewpoint over the chicken farm, a pig farm comes into sight and a track is seen coming from the west and bordering an orchard. The corner of the track is at 36 528948 E, 38 74827 N, and on the shoulder of the hill above this corner is an excavation into the hillside that exposes pillows of olivine-rich lava and orange-brown interstitial chert-rich sediment (Fig. 3.13: P).

There are many other features of interest in the area, which are well worth exploring if there is time. It should be noted that there may be a link between olivine-phyric lavas, such as those seen here, and late ultramafic plutons that occur in the mantle and plutonic sequences.

Sulphide deposits and umber

The iron, copper and zinc sulphide deposits associated with the volcanic sequence of the Troodos ophiolite are world famous as the originals of the Cyprus type of volcanic-hosted massive sulphide deposits. In Cyprus these deposits cluster in five major orefields (see Fig. 2.2) and are made up of lenses that usually contain no more than a few million tonnes of massive sulphide. The sulphide is called "massive" because it has little else in it apart from pyrite. These deposits have many features in common with those seen forming today around the black-smoker hot springs at seafloor spreading centres, and the Cyprus deposits have, therefore, contributed significantly to understanding the processes of submarine hydrothermal circulation and associated mineralization.

A general model for hydrothermal circulation at a spreading centre is set out in Figure 3.14. The process is based on a type of convection that

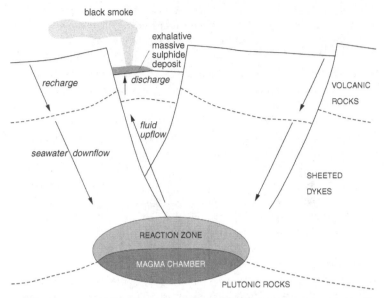

Figure 3.14 Overall structure of an active mid-ocean ridge black-smoker system, showing the circulation pattern of the convecting sea water and hydrothermal fluid, the heat source in the magma chamber, the hydrothermal reaction zone, the sulphide deposit on the sea floor and the plume of black smoke.

begins with cold sea water flowing down through pores in rocks of the oceanic crust. At depth and close to a magma chamber or other heat source, such as a newly crystallized pluton, there is a reaction zone where sea water becomes heated and chemically modified to hydrothermal fluid. This fluid then rises rapidly through a narrow upflow zone to be discharged through black-smoker hydrothermal vents on the sea floor. Formation of a Cyprus-type deposit of several million tonnes of sulphide would have required tens of cubic kilometres of hydrothermal fluid and a very large amount of energy.

Modern black smokers

In the modern oceans, black-smoker vent fields have been found at many sites along the mid-ocean ridges and in back-arc basins, close to spreading axes and associated with young volcanic rocks. Within these vent fields, which are a few hundred metres across, there may be many black-smoker chimneys made of sulphide, from which hydrothermal fluid emerges at 350–400° C. This fluid is a solution that is acidic, poor in oxygen and rich in dissolved hydrogen sulphide, and it contains high concentrations of iron, copper, zinc and manganese. As the emerging fluid mixes with cold sea water at the sea floor, it precipitates a cloud of black particles of iron sulphides (the black smoke), which rises 100–200 m above the sea floor before being swept away by ocean currents. In the oxygen-rich sea water, the particles of sulphide in the black smoke are oxidized to iron oxide, which then takes up manganese and other elements from the water. The resulting cloud of iron and manganese oxide gradually settles to form a layer of metalliferous sediment surrounding the vent field. Around the vents there is also a pile of sulphide that is made of collapsed pieces of chimney and sulphide precipitated at or just below the sea floor.

Perhaps one of the most astonishing discoveries in modern science was that black-smoker vent fields host biological communities, within which the productivity per square metre is as great as that in a tropical rainforest. The organisms in these communities include tube worms, clams, mussels, shrimp, crabs and fish. The communities are supported by chemosynthetic microbes that depend for life on the chemical energy from the oxidation of vent fluid by sea water.

Cyprus-type hydrothermal deposits
The Troodos ophiolite preserves all of the components found within a hydrothermal system. Unfortunately, most of the sulphides were mined out by extensive open-pit mining in the twentieth century and, since then, many of the pits have been degraded by weathering. Nevertheless, from the descriptions of the deposits made during mining and from the few pits that may be usefully and safely visited, a clear picture does exist of the idealized Cyprus-type volcanogenic massive sulphide deposit (Fig. 3.15). At its heart is a lens of massive sulphide, which is usually made up of more than 90 per cent pyrite, and is up to several hundred metres across and up to tens of metres thick. The pyrite is in some places coarsely crystalline and yellow, with good crystal shapes, but more often it has a fine grain and is black to silvery with colloform textures. Within the massive sulphide it is possible to find fragments of black-smoker chimneys and fossils of tube worms and gastropods that once lived around these chimneys. The fine-grain pyrite and the chimneys are interpreted as primary products, whereas the coarse crystalline pyrite seems to have formed by recrystallization of primary pyrite as the sulphide pile grew and hot hydrothermal fluid percolated through it. Chalcopyrite and sphalerite form only minor components of most of the massive deposits.

Immediately beneath the massive pyritic lens is a unit known as the stockwork zone (Fig. 3.15). It is defined by a complex network of veins of pyrite and quartz cutting through basalt, which has been altered by the passage of hot hydrothermal fluid, and it represents the feeder zone for the overlying lens. Fluid rising from deeper in the oceanic crust was probably focused along structural conduits, such as the major fault that is usually

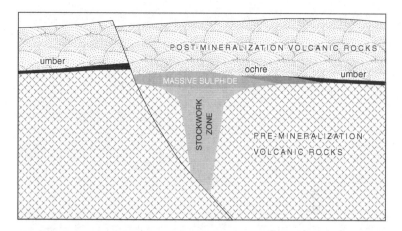

Figure 3.15 Structure of an idealized Cyprus-type massive sulphide deposit, showing the hydrothermal feeder (stockwork) zone in pre-mineralization volcanic rocks, the exhalative massive sulphide deposit with its associated ochre and umber, and the post-mineralization volcanic rocks. The diagram implies that the sulphides were deposited in a half-graben and that volcanism and fault movement continued after sulphide formation.

found close to a sulphide deposit. This fluid would have risen from the deepest part of the hydrothermal system – the reaction zone – which is superbly preserved in Cyprus in the form of epidosites in the sheeted dykes (stops 3.7, 3.8). It is here that metals were stripped from the igneous rocks, as they were altered by the hydrothermal fluids.

Almost all of the sulphide deposits were covered by later lavas that were not altered by hot hydrothermal fluids. These lavas therefore postdate the hydrothermal activity and may have been fed by unmineralized dykes that in places intrude mineralized lavas (see stop 3.5). These relationships suggest that most of the vent fields must have been active very close to the fissures from which the lavas were erupted.

Associated with the sulphide deposits are units of very fine-grain brown metalliferous sediment known as umber, made of the iron and manganese oxides of the black smoke. The thickest units of umber occur immediately above the uppermost lava flows, filling depressions in the lava surface, where they may reach thicknesses of 35 m. Locally these umbers grade upwards into deepwater radiolarian mudstones and marl-rich sediments (Fig. 3.16). Taken together, these sedimentary rocks make up the Perapedhi Formation, which is the oldest unit in the circum-Troodos sedimentary succession. Umber also occurs within the volcanic rocks, as thin layers between lava flows and filling gaps between individual pillows and between clasts in submarine talus. Bright-red jasper occurs where umber was altered by hot hydrothermal fluid.

Figure 3.16 Skouriotissa mine (stop 7.3). **(a)** The eastern end of Foucassa pit in 1994. Altered and mineralized volcanic rocks to the left and centre of the photograph are faulted against dark unmineralized lavas to the right. These rocks are overlain by ochre, umber and white marl, and the fault displaces them. The fault may have originated early during mineralization of the volcanic rocks. The massive sulphide lens that once existed has been completely removed by mining. **(b)** Part of a cutting in Phoenix pit, which exposes a lens of massive pyrite overlain by umber and underlain by fractured altered pillow lava. The height of the section is about 5 m.

After the massive sulphide deposits formed close to the spreading axis, and after umber had accumulated on the top of the lavas, there was renewed hydrothermal circulation in the ophiolitic crust. This took place at low temperatures and was associated with silicification of the umbers and the formation of gold deposits within the umbers or within the uppermost lavas, or both.

Many of the sulphide deposits record a history of weathering on the sea floor and upon exposure above sea level. Submarine weathering and reworking of the derived sediment produced well bedded ochre, a mixture of hydrated iron oxides with a very low content of manganese. These colourful (often yellow) banded rocks should not be confused with umber and they are found resting directly on the top or the sides of a lens of massive pyrite.

During the Quaternary, as the ophiolite was uplifted above sea level and eroded, the sulphide deposits were exposed. Once they arrived above the water table in the Mediterranean climate, microbial oxidation of the pyrite began. Pyrite was converted to a mixture of iron oxides and hydroxides, iron oxyhydroxysulphates and sulphuric acid. The acidic water further weathered both the sulphide and the rocks enclosing it, often converting the stockwork zone into a silica sponge, a rock made up mostly of a network of porous quartz that once cemented the now-dissolved pyrite grains. The oxidation of the sulphides led to the formation of distinctive deep yellow, orange, red and brown iron-bearing caps at the tops of the sulphide orebodies. These gossan caps have turned out in places to be enriched in gold, and the ancients appear to have used them to find the underlying sulphide orebodies.

Most of the mineralized zones in the volcanic sequence have been mined and subsequently weathered to the point where many key features have been obliterated. The four stops presented permit meaningful examination of hydrothermal systems. Stops 3.16 to 3.18 are close together and it is possible to walk from one to the other in sequence. This is best done if transport drops off at stop 3.16 and picks up at stop 3.18. Other relevant stops are 3.7, 3.8, 7.1, 7.3 and 7.12. Caution must always be exercised when visiting any mined area.

Stop 3.15 Sulphide mineralization, umber, volcanic rocks and acid drainage at the Mathiatis mine From the E103 between the villages of Agia Varvara and Mathiatis, the large spoil heaps of the former Mathiatis mine are clearly visible. To reach the mine, leave the A1 at junction 8, drive southwest through Agia Varvara and then a cutting through a small ridge composed of lavas, and park on a large flat barren area to the east (left) of the road. From the other direction, the site is about 2 km from the village of Mathiatis, and the parking area is on the right before the roadcut. The starting point for the stop is at the northwestern end of the open pit, at the gate across the track that leads down into the pit (36 531583 E, 38 70656 N). Taking care of the rim, obtain a good view over the pit.

91

The sulphide deposits at the Mathiatis mine were worked by the ancients, but the modern mine did not open until the 1930s, when precious metals were exploited. Following a period of underground mining, large-scale open-pit working of the medium-size massive sulphide deposit took place between 1965 and the mid-1980s, when the mine closed. During this time, about 3 million tonnes of low-grade copper ore was removed from the site. After the end of mining, the pit partly filled with surface water, which is today pumped out and used for irrigating nearby fields, olive groves in particular. These waters leave precipitates of iron oxides and sulphates on and in the soil, indicating that they are acidic and metal-bearing. Perhaps surprisingly, the combination of acidic water and alkaline soil, derived from the weathering of underlying lavas, makes for good growing conditions. The water in the mine is in fact some of the least contaminated mine-pit water in Cyprus, and it has relatively low conductivity and moderate acidity. Many people comment on the colour of the water, which varies from green to dark reddish-brown to copper-sulphate blue. These colours are unlikely to be controlled solely by metals in the water, but also by temperature, oxygen content, acidity, biological activity, rainfall and the concentration of very fine sediment.

At first glance the geology of the pit appears simple. To the southeast, the opposite side exposes an orange zone of weathered stockwork, whereas the northeastern and southwestern walls of the pit expose lavas unaltered by hydrothermal fluids. From the viewpoint, walk a half circle down into the pit and then look to the northwest, back towards a steep wall of pillow lavas below the viewpoint. The elongate massive sulphide lens originally ran overhead and plunged down into the water. The last remaining piece of this sulphide in the pit lies below water beneath the steep wall of pillows. These pillows were erupted after the ore formed and are unaltered. The place where you are standing is in the stockwork zone that underlay the sulphide lens, and on either side there are moderately altered lavas that lay below the level of this lens, but were outside the reach of the hot hydrothermal fluid coming up the stockwork zone. Looking back to the northwest and the wall of pillow lavas, there appear to be two steep faults. They cannot be traced far into the pit with any certainty and it is not clear which one, if either, was the main fault along which hydrothermal fluids were channelled. The faults appear to be overlain and truncated by umber, suggesting that they definitely pre-date mineralization, but this umber could have been derived from venting farther afield.

Walk farther around and down onto the next level to where there are excellent blocks of stockwork. Thick veins of pyrite and yellowish quartz cut grey, silicified, hydrothermally altered basalt. In places there are inclusions of jasper in the stockwork resulting from hydrothermal alteration of umber trapped between pillows. Judging by crosscutting relations, the silicification was the last hydrothermal event, because the youngest veins that cut the pyritic veins are composed entirely of quartz. At Mathiatis the stockwork is more highly silicified than in most other Cyprus deposits.

During the walk back up to the gate it is worth examining the transition from the intensely altered and silicified stockwork, containing red jasper, through the moderately altered lavas, to the relatively unaltered lavas in which the umber between the pillows is brown. The lava section contains some excellent pillows and sheetflows. Note that the roadway is surfaced with black slag from the ancient local copper workings.

Back at the gate, follow the track as it climbs around the northern rim of the pit. Along the way are blocks of massive sulphide from the last stages of mining. This waste is oxidizing, as is evident from the extremely pungent smell of sulphur dioxide and the variety of coloured oxidation products. During rain, some of these minerals dissolve to produce highly acidic metal- and sulphate-rich drainage waters. A range of colourful sulphate minerals precipitate from this acid drainage, some of which have an algal form suggestive of biomineralization (see Fig. 7.3). During hot weather, within hours or days of rainfall, it is amazing to see the growth of efflorescent minerals on the porous pyritic waste and in drainage channels.

A little farther up the track, take the first turning left onto another track heading northwards, along which there are huge blocks of grey pyrite undergoing oxidation. As the track leaves the waste pile, head towards the hillock to the north-northwest by skirting around a field. On the hillside, as the field flattens, there are exploration pits into subhorizontally bedded umber. Farther up on the hilltop there is a contact between umber and a vertical face of a brecciated columnar-jointed flow (36 531898 E, 38 70747 N). The breccia clearly pre-dates the umber and may represent a fault scarp or the steep wall of what was originally a submarine depression. Across the field to the west is another hillock of lava, but the field in between is underlain by umber. This relationship of umber filling in the depressions originally present on the lava surface is common and is clearly seen in the distance to the north and northwest from the northwestern end of the exposed breccia/umber contact. The umber along the contact is quite distinct from that in the exploration pits, as it appears to be deformed and veined, and it is steeply dipping. The dip is clearly defined by black nodules that look like manganese nodules, but are umber that has been silicified. These nodules appear boudinaged and may indeed be deformed, probably as a result of compaction (see stop 3.16). The silicification of the umber probably took place during a late phase of low-temperature hydrothermal circulation (refer to stops 3.16 and 3.18).

Return to the track and continue along it in an easterly direction in order to examine the umber/lava contact in pits left by the removal of umber. In the third and largest pit the umber lies above lavas and below pinkish marl of the Lower Lefkara Formation. Beyond this pit it is possible to see plenty of outcrops of volcanic rocks, especially pillow lavas, that in places are intruded by near-vertical dykes possessing excellent chilled margins.

Stop 3.16 Half-graben filled with umber and mudstone of the Perapedhi Formation near the Theotokos monastery east of Kampia   The area around the Theotokos monastery is one of the most instructive in the upper part of the ophiolite, especially for its umber deposits and its faulting (Fig. 3.17). The outcrops of interest lie a short distance south of the monastery and are reached from the road between the villages of Analiontas and Kampia. From the centre of Kampia, travel about 1 km east and park on the right (south) side of the road on open ground just beyond an almond orchard, and just before a track running steeply down hill to the right (36 524130 E, 38 74205 N). The track leaves the road at the point where the monastery first comes into view. Do not try to drive down the track. Coming from the east, the parking place is about 3 km from Analiontas, after the two roads leading to the monastery.

From the parking area, follow the track down hill for about 400 m and then take a well defined track to the left. The track winds around a nose of rock running down from the north. Stop at the crest of the nose and look north. The nose is made of pillow lava, and on top of the hill are some large brown boulders. Walk north up the crest of the nose, veering slightly left on the way. On the western slope of the nose the pillows are covered by well bedded dark-brown umber, which dips steeply westwards. Lift a piece of the umber to discover that it has a very low density and then lick it gently to allow the tongue to stick, this indicating that the umber is highly porous. Now look very closely at a section perpendicular to the bedding to find very narrow white veins that run across the bedding and are crinkled into microfolds. The crinkling shows that the umber has been compacted since the veins were formed and must, therefore, have originally been much more porous than it is now. From the geometry of the microfolds, it appears that compaction has reduced the umber to a half or a third of its thickness since the veins formed, which must have been some time after the umber was deposited.

Now climb up to the brown boulders on top of the hill (36 524612 E, 38 73882 N). These are highly silicified (about 90% silica), but, when the silica content is subtracted, the boulders have the composition of umber. This relationship might be explained if the boulders represent umber that

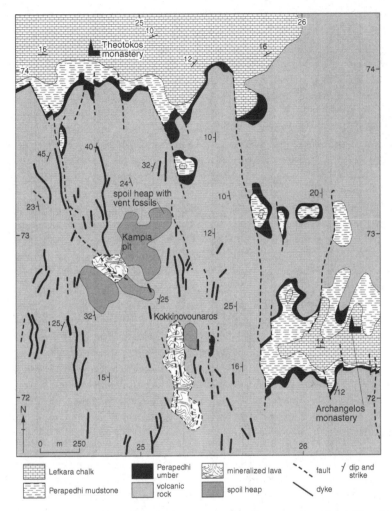

Figure 3.17 Geological map of the area between Theotokos monastery (stop 3.16) and Kokkinovounaros (stop 3.18) simplified from the unpublished maps by Richard Thompson and Rex Taylor. The numbers around the border are from the WGS84 UTM grid, where, for example, easting 25, northing 74 corresponds to the full grid reference 36 525000 E, 38 74000 N.

was silicified before it was compacted, when the porosity of the umber was close to 90 per cent. If so, the umber on the western slope of the nose may have been compacted to perhaps 20 per cent of its original thickness.

From the brown boulders, look around. To the north the monastery stands on top of a chalk cliff, the lowest part of the Lefkara chalks that once were several hundred metres thick above this place. Down to the east is

95

more umber, but it sits 20–30 m lower than the umber on the hilltop. Above the umber, as can be seen by clambering around, is a 20 m-thick unit of pink-to-grey siliceous mudstone, the top of which is at the same level as the brown boulders. The mudstone contains tiny white remains of radiolarians, which are best seen by examining a wet surface with a hand lens. These mudstones and the umber belong to the Perapedhi Formation.

From the hilltop, clamber towards the monastery along the steep slope running down to the east. At first there is an outcrop of greenish highly weathered basaltic hyaloclastite, and a bit farther on there is a ledge with umber on it, half-way down the slope. The umber has a curved fault surface on it, which is decorated with the pink manganese-carbonate mineral rhodochrosite.

The observations fit together by considering the structure, which is a half-graben – one of at least three small half-graben running north–south beneath the monastery. They are bounded by faults on the west side that throw down to the east (Fig. 3.17). The faults formed very early and the half-graben filled with umber and then with siliceous mudstone. At some early stage, before the umber had compacted, warm hydrothermal solutions came up the fault in one place and silicified the umber. It is interesting to wonder whether the faults went on moving as the umber accumulated. Certainly, the mudstones were laid down after faulting was complete. Eventually the half-graben and their fill were covered by Lefkara chalk, about 25 million years after the ophiolitic crust had formed.

From here it is possible to make a transect across country, first to the pit at the Kampia mine (stop 3.17) and then to Kokkinovounaros (stop 3.18). The geology along this transect is shown in Figure 3.17. Ideally, transport should be available at stop 3.18.

Stop 3.17 Massive sulphide, vent fossils and structures at the Kampia mine west of Analiontas   In the centre of the village of Analiontas, turn west onto a narrow metalled road immediately south of the bridge over the river and follow this road to the west, keeping the river on the right. The road turns into a gravel track that is passable with care. Keep right at a major fork and soon the spoil heaps of the former Kampia mine become visible ahead. Stop by a grove of eucalyptus trees close to the first spoil heaps (36 525127 E, 38 73210 N). Those walking across country from stop 3.16 will be able to see to the south the spoil heaps of the mine (Fig. 3.17). Aim for the easternmost spoil heaps and the grove of eucalyptus trees near them. The outcrops of dipping sheetflows along the way are worth examining.

The spoil heaps contain blocks of massive pyrite, some with very fine grain that are silvery, and others that are coarser and yellow. Careful searching will reveal good examples of a wide range of ore types, exactly

Figure 3.18 Vent fossils from Cyprus massive sulphide deposits: **(a)** coiled tube worm; **(b)** gastropod. All material is sulphide because the original tube and shell have been replaced. Each image is about 3 mm across. (Both images kindly supplied by Crispin Little)

the same as those that develop at black smokers on the modern sea floor. Patient examination of blocks of the silvery sulphide with a hand lens will reveal that they contain tiny straight, curved or coiled tubes (1–2 mm in diameter), which are the fossils of tube worms (Fig. 3.18a). These fossils are very similar to tube worms known to exist around black smokers today and they demonstrate conclusively that their hosting sulphide formed at black smokers on the sea floor. In other blocks of the silvery sulphide there are tiny fossil gastropods (Fig. 3.18b).

Now continue about 300 m farther west along the track, to where it bends first left and then right, and park by a fenced enclosure with a concrete shed in it (36 524914 E, 38 73064 N). Walk up the old track on the right of the enclosure to a fence at the top that surrounds the pit. Do not go through the gate or cross the fence, as the pit is in an extremely dangerous state. Walk left (south) to a spoil heap containing varied types of sulphide and blocks of chert coloured by bluish-green celadonite. Continue to a viewpoint at the southern end of the pit, from where the structure of the pit becomes clear. There is a major north–south-trending fault on the western side of the pit, and the lavas west of it are dark and unaltered, whereas those east of it are grey and are part of the stockwork that underlay the massive sulphide lens. Immediately east of the fault is a unit of talus in which the voids between the clasts of basalt are filled with umber. The umber has been altered to jasper and the basalt has been silicified and pyritized during the hydrothermal activity. At the far end of the pit there is more stockwork visible, which was profoundly weathered near the old land surface. At the top of the hill above the northeast corner of the pit there are lavas that postdate the sulphide deposition and they are therefore not hydrothermally altered.

Stop 3.18 Low-temperature hydrothermal mineralization and sulphide weathering on Kokkinovounaros southwest of Analiontas The Kokkinovounaros locality (see Fig. 3.17) shows both the results of off-axis hydrothermal circulation associated with gold deposition and the effects of weathering of sulphides as they were uplifted above sea level in the Quaternary. The deposit is on a conspicuous red-weathering hilltop that lies southeast of the Kampia mine (stop 3.17). About 200 m south of the village of Analiontas, on the road to Lythrodontas, turn west onto the road signposted Moni Arkhangelou (the Archangelos monastery). The metalled road heads south past the village football pitch and up hill to pass the monastery at the crest. Here the road turns west and heads directly for the red-topped hill of Kokkinovounaros. Park where the road turns south, close to the top of the hill, and climb up a track to an open space near the north end of the hill, where there is an adit driven into the rock (36 525236 E, 38 72405 N). Those walking across country from stop 3.17 should head towards a point somewhat east of the red-topped hill, and reach a ridge that runs up from a dirt track to the summit from the north.

On the hill the effects of the late low-temperature hydrothermal circulation and the Quaternary weathering are closely superimposed. Originally the deposit was composed of seafloor lavas overlain by umber and cut by a north–south fault (Fig. 3.19). Late-stage hydrothermal solutions rose along the fault, altered the lavas and umber, and precipitated pyrite, quartz and gold within the lavas. The deposit was mined for gold some decades ago. The zone of mineralization is elongate along a north–south fault that follows the crest of the hill, along which most of the mining took place, and is now represented by an elongate north–south pit. The most striking feature of the deposit is its range of colours: pure white, pink, scarlet, black, browns and bright yellows. The coloration reflects the variety of iron-based alteration minerals, which are the product of extreme weathering during the Quaternary of the disseminated pyrite within the now-altered lavas that make up the ore. As the pyrite was exposed, it became oxidized and this generated acid solutions that percolated through the rocks and leached, oxidized and weathered them intensely. The unmineralized lavas in the area are noticeably much less weathered.

Start in the open space in front of the man-size adit that was driven into pink hydrothermally altered and subsequently weathered basalt (Fig. 3.19). Pillow structures can be seen in a cut that links the open space with the central pit about 30 m left of the adit. Heaps of cobbles in the open space and on a terrace to the east of the adit are the last ore to be brought out of the mine and are similar to the pink basalt that surrounds the adit. Some of the rocks in these heaps contain traces of invisible gold. A few metres above the adit is a unit of well bedded black and yellow altered umber, which can be reached by clambering up onto the terrace above the adit. It is not as silicified as the brown boulders at stop 3.16 and must have been impregnated with pyrite, to judge by its weathering colour.

Next, walk through the cut into the north–south pit in the centre of the deposit. Here it is possible to find pure white silica sponge (silicified lava that has been leached of everything but quartz during weathering), yellow jarosite (a sulphate mineral formed during weathering) and much more.

The energetic may wish to look for small exposures of unweathered altered basalt veined by pyrite and quartz, which crop out at both the north and south ends of the top of the hill. These must represent the nature of the deposit before it was weathered in the Quaternary. About half-way along the top of the hill, on its eastern side, is another outcrop of altered umber, which is highly silicified.

Figure 3.19 Altered umber overlying volcanic rocks in the mined area at stop 3.18. The pile of cobbles represents the ore once mined for gold. The open pit lies on the left of the photograph.

Chapter 4

The Troodos ophiolite: complex faulting

Introduction

The main body of the Troodos ophiolite can be divided into two geologi-
cally and geographically distinct areas: the Troodos massif and the Limassol Forest (see Fig. 1.2). The massif commands by far the largest area and
is examined in detail in Chapters 2 and 3. The present chapter is concerned
with the much smaller southeastern lobe of the ophiolite, which lies to the
north and northeast of Lemesos. This part of the ophiolite is referred to as
the Limassol Forest because most of it lies within the mountainous region
of the forest. The lithologies in this area are all essentially ophiolitic, but
their interrelationships are quite different from those in the rest of the
ophiolite because they have been extensively reorganized by faulting. The
northern limit of the Limassol Forest is marked by the prominent east–
west-trending Arakapas Valley (Fig. 4.1), which contains the Arakapas
fault zone, and is considered by many to be the surface expression of a fossil transform fault. In modern ocean basins these faults offset segments of
mid-ocean ridges, most spectacularly in the equatorial Atlantic, but rarely
are they preserved in ophiolites. It is therefore not surprising that the
Arakapas Valley and adjacent areas are of considerable geological importance.

The Arakapas Valley and Limassol Forest

The Arakapas fault zone is about 1 km wide and consists of fault strands
interpreted as strike-slip faults that separate and link along strike (Fig. 4.2).
Its east–west trend is at right angles to the generally north–south trend of
the sheeted dykes farther to the north. The surface expression of the fault
zone formed a linear depression on the ancient sea floor, as shown by its
infill of later mafic lava flows that are intercalated with a variety of clastic

Figure 4.1 View north from the E109 south of Eptagoneia, showing the village on the northern margin of the east–west-trending Arakapas Valley. The northern side of the valley is marked by the steeply rising mass of sheeted dykes of the main part of the Troodos massif. The foreground is underlain by sheeted dykes also, but the valley bottom between these two dyke units contains volcanic and sedimentary rocks, and the east–west-trending Arakapas fault zone (see Fig. 4.2).

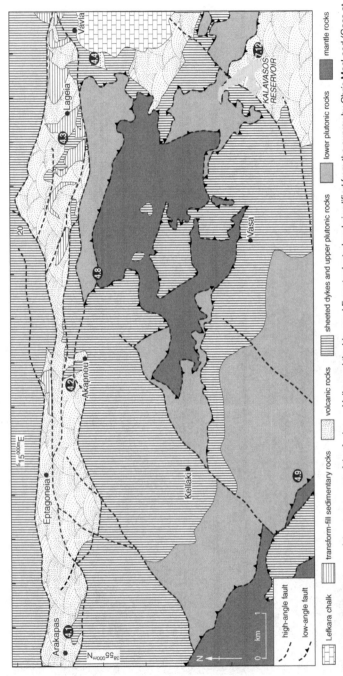

Figure 4.2 Geological map of the eastern part of the Arakapas Valley and the Limassol Forest adapted and simplified from the survey by Chris MacLeod (Gass et al. 1994). The Arakapas fault zone runs east–west across the top of the map and is marked by interlacing faults and an infill of volcanic and sedimentary rocks. Numbers refer to stops; stops 4.6 and 4.7 lie just off the western side of the map. The numbers around the border are from the WGS84 UTM grid.

sediments derived from the valley sides. These infilling lavas and clastic sediments are deformed only weakly, whereas the pelagic sediments that overlie them are undeformed. This evidence led to the suggestion that the Arakapas Valley does indeed represent a fossil transform fault. The infill is traceable well over 20 km farther west from the end of the Limassol Forest area, showing that the fault zone must form the southern boundary of the ophiolite over this distance, giving a total length of the fault zone in excess of 60 km. The infill lavas have compositions and textures that show strong affinities with the highly magnesian and siliceous volcanic rocks (boninites) of island arcs, and are distinct from the lavas in the ophiolite north of the fault zone.

South of the Arakapas Valley, the Limassol Forest hosts all of the lithological units of the ophiolite that are seen north of the Arakapas fault zone, as well as the volcanic and volcaniclastic units that are found along this fault zone. However, the lithological units within the Limassol Forest do not form a systematic layer-cake structure dipping consistently in one direction, as they do in the rest of the ophiolite to the north of the Arakapas Valley. They are instead distributed apparently more randomly and large sections of the lithostratigraphy may be missing between units that form part of the normal stratigraphy north of the fault zone. For example, lavas are not just in contact with sheeted dykes, but also with gabbro and serpentinite. Detailed mapping has shown that this is the result of complex faulting, much of it with a low angle of dip. Omission of lithological units suggests that the faulting was extensional, and the low dip of the extensional faults gives rise to a complex outcrop pattern and to sinuous contacts between the units in plan view (Fig. 4.2).

Broadly, the Limassol Forest comprises an east–west core of serpentinite and gabbro that is bounded and covered by fault slices of sheeted dykes and volcanic units. In the west, serpentinite dominates and there is evidence that large blocks of this mantle rock, and others of crustal rocks, have been emplaced during their tectonic uplift and rotation. Later magmatism produced plutons of mafic and ultramafic (mainly wehrlitic) compositions, with related dykes that acted as feeders to lavas with the same distinct boninitic characteristics as the infill lavas in the Arakapas Valley. In the south and east, gabbro, sheeted dykes and lavas are more common, and later intrusive wehrlitic bodies are again found. Within the large bodies of serpentinite there are vertical east–west shear zones, parallel to the Arakapas fault zone, filled with highly sheared serpentinite. These shears do not penetrate into the gabbros, sheeted dykes or lavas.

Taken together, the field evidence from the Limassol Forest shows that the tectonics of the area are characterized by two major low-angle extensional faults, which separate serpentinite below from a variety of rock types above; these higher units are themselves divided into slices by more low-angle faults. Mapping shows that the deeper structural units in the hanging-wall blocks lie broadly west of the shallower structural units, and that some of these hanging-wall blocks have been rotated through large angles. The vertical east–west shear zones in the serpentinite may be strands of the transform fault or they may be strike-slip relay zones on the low-angle extensional faults. Some late fault movements in the Limassol Forest may be associated with the anti-clockwise rotation and obduction of the ophiolite. Other faults can be firmly timed with a phase of compression linked with Middle Miocene uplift of the Limassol Forest area (see stops 6.2, 6.6, 6.11).

Field relations show that both the extensional faulting and the serpentinization of the harzburgite were mostly complete at a very early stage. Serpentinite was exposed on the sea floor before or during the eruption of late (boninitic) lavas in the southern part of the Limassol Forest, and serpentinization and deformation pre-dated the intrusion of late dykes and plutons. Consequently, extensional faulting took place during the formation of the ophiolite.

Pulling all of this information together into a coherent story is difficult, but there are ways to explain the extensional faulting. It may have happened in a ridge-transform inside-corner environment as part of the spreading process (Fig. 4.3a) or in a broad (20–30 km-wide) zone of transform-deformed crust within an active transform-fault zone (Fig. 4.3b). A third possibility is that extensional faulting may have happened over a prolonged period following crustal accretion as part of a process related to the supra-subduction zone environment of the ophiolite. Each explanation makes specific predictions about the field relations that ought to be visible. For example, the third explanation is made less likely by the dating of the extensional faulting relative to the later intrusions summarized above.

The Limassol Forest and Arakapas Valley preserve a complex geological history of seafloor spreading, transform faulting, extensional faulting and magmatism at the time of ophiolite formation, as well as deformation related to the anti-clockwise rotation of the ophiolite in the uppermost Cretaceous. To unravel this story takes more than the few stops presented here, but these stops demonstrate a wide variety of field relationships and all should be visited to appreciate the geology of this rare natural laboratory. Stops 4.1 to 4.4 provide a west–east transect that examines the Arakapas fault zone, and stops 4.5 to 4.9 explore the Limassol Forest, the volcanic rocks of which are examined at stop 7.12 (Fig. 4.2). The majority of the stops may be combined into a day-long excursion that begins north of Lemesos and follows the sequence 4.5, 4.6, 4.7, 4.1, 4.2, 4.8, 4.3 and 4.4.

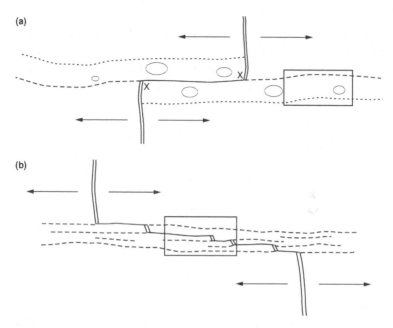

Figure 4.3 Two models for the formation of the Limassol Forest in a spreading environment (indicated by arrows) offset by transform faults. **(a)** At a ridge–transform inside corner (X), where the strip of crust produced at the inside corner (bounded by the dashed lines) has been subjected to extreme extension symbolized by the ellipses. **(b)** In a broad transform deformation zone in a complex multi-stranded transform system. In both cases the area in Figure 4.2 is represented by the rectangle.

Stop 4.1 The Arakapas Valley and volcanic rocks at Arakapas   The village of Arakapas is partly bypassed by a new road linking the F128, coming north from Akrounta, to the F129 coming west from Eptagoneia. On the southern side of the bypass there is a cutting, which is about 150 m south of the old church (with a steep tiled roof) in the village and about 100 m west of the junction between the bypass and the road into the village (36 510555 E, 38 55633 N); park at this junction.

What is most obvious about the village is that it sits at the bottom of the east–west-trending Arakapas Valley. The roadcut exposes pillow lavas cut by a few dykes, and the presence of some fresh basaltic glass suggests that these rocks are not significantly metamorphosed, unlike those at stop 4.2. The outcrop is cut by one prominent near-vertical fault that runs almost parallel to the road. The amount of deformation is small and the fault is stained red by iron oxides that probably formed during low-temperature fluid circulation. The lavas were apparently erupted into the trough of the Arakapas fault zone after most of the slip on the transform fault had ceased.

Stop 4.2 Deformed and metamorphosed mafic rocks in the Arakapas fault zone at Akapnou The village of Akapnou lies 100 m or so from the F124/F114 road between Eptagoneia to the west and Ora to the north. About 300 m north of the entrance to Akapnou, towards Ora, there is a cutting on the eastern side of the road (36 517100 E, 38 55565 N), and there are plenty of places to park along the road.

The cutting exposes brecciated dykes and some pillow lavas, all show-ing an overall green colour that indicates metamorphism to greenschist facies. Parts of the outcrop have been mineralized, so that abundant cubes of pyrite have grown within the metamorphic rocks. The exposure is cut by a variety of faults in different orientations, but none of these are vertical and east–west as the simple transform model would imply, suggesting that deformation here has been complex. The exposure is interpreted as a highly deformed and metamorphosed part of the basement of the Araka-pas fault zone. The volcanic rocks cropping out at stop 4.1 would have accumulated on top of this type of basement.

Stop 4.3 Sedimentary and volcanic fill in the Arakapas fault zone west of Lageia A few hundred metres west of the village of Lageia there is a straight sec-tion of the F112 to Ora that cuts through a superb sequence of eastward-dipping red and brown sedimentary rocks (Fig. 4.4). To examine these, park on the flat area at the eastern end of the cuttings (36 522400 E, 38 56088 N), but begin observation at their western ends.

The lowest unit in the sequence is a coarse sedimentary breccia. Clasts

Figure 4.4 Sedimentary and volcanic rocks in the Arakapas fault zone (stop 4.3). A breccia unit lies below a unit of turbidites capped by a basaltic lava flow.

are poorly sorted and they range in size from a few centimetres to several metres. They include blocks of lava, dolerite, rare gabbro and clastic sedimentary rock. Two large flat slabs of clastic sedimentary rock, a metre or more thick and several metres wide, lie parallel to the bedding in the middle of the breccia unit on the north side of the road, and probably show the orientation of the bedding here. These slabs are made of graded units of sandstone, changing from coarse to fine grain downwards so that they appear to be inverted. Red mud seems to have been forced up from below into the sandstone. The breccia unit must have accumulated very close to a source that included not only outcrops of volcanic and dyke rocks, but also units of consolidated clastic sediments that could be reworked into the breccia as large clasts.

Above the sedimentary breccia there is a rapid transition to a unit of alternating red mudstone and grey sandstone. This unit is a series of turbidites, each one tens of centimetres to a metre or so thick. Upward grading between the sandstone and the mudstone is clearly demonstrable, as are other turbiditic characteristics. The sands have the same basaltic lithology as the larger clasts in the unit below. The mudstones are dark red and apparently much more iron rich, and probably represent reworked sediment from umber.

Above the unit of turbidites is a uniform brown basalt, probably a single lava flow. Between this flow and the underlying turbiditic mudstone is a brown shale, which shows deformation features indicative of fault movement. Further deformation is recorded in the first turbidite unit beneath the basalt. Here there is a bedding-parallel thrust fault showing small-scale repetition of thin turbidite units within the fault zone. Note that the conglomerate at the eastern end of the roadcut is not a seafloor feature but a Quaternary boulder fan.

The sedimentary and volcanic sequence exposed in the roadcuts must have accumulated on the floor of a valley, presumably one with a very steep side, at least at the time when the lowest unit was being deposited. Yet the material being shed into this valley apparently included clasts of sedimentary rock that must have formed in a topographical low. To explain this situation, it is informative to look at modern transform faults, such as the Clipperton fault in the Pacific. Here the active transform is sinuous, not straight, so that some of the sections are in compression and others are in extension (Fig. 4.5). Along such faults, therefore, small extensional basins are separated by compression-generated highs, and it is likely sediment that previously accumulated in a basin may, as the fault slips, become elevated in a compressional section and be able to supply clasts to neighbouring basins. Certainly, the nature of the sedimentary rocks outside Lageia is consistent with their having formed as part of the

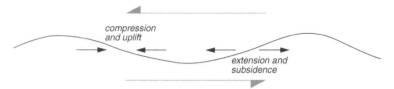

Figure 4.5 Plan view of a model for compression (uplift) and extension (subsidence) along a sinuous transform fault.

fill of a transform-fault trough, with the lava flow also being part of this fill and equivalent to the lavas seen at Arakapas (see stop 4.1).

Stop 4.4 Sedimentary fill of the Arakapas fault zone west of Vavla   On the F112 between Lageia (to the west) and Choirokoitia (to the southeast) there are two turnoffs to the north into Vavla. A short distance east of the western turnoff is the start of a track heading south, where it is possible to park (36 524068 E, 38 55548 N). To the west are dark basaltic rocks, which contrast with the white Lefkara chalk to the east. Head south on the track.

To the west of the track, and up hill from it, is a unit of basaltic breccia. To the east of the track, and down hill from it, is a remarkable sequence of sedimentary rocks. Above the breccia is a unit of red siliceous mudstone of the Perapedhi Formation, showing chevron folds verging towards the east. This folding probably arose from slumping at the time of accumulation of the sediments. Above these folded rocks is a poorly exposed unit of greyish bentonite, which is the lateral equivalent of the Kannaviou Formation in western Cyprus and the Moni Formation south of the Limassol Forest. Above the bentonite is Lefkara chalk.

The Perapedhi mudstones and the overlying bentonite are locally thickened here near Vavla. This fact and also the slump folding in the mudstones probably reflect the continuing presence of a valley along the line of the Arakapas fault zone long after the fault had become inactive.

Stop 4.5 Lava flows and volcaniclastic sedimentary rocks in the Limassol Forest north of Akrounta   From the F128 between Akrounta to the south and Arakapas to the north, about 1 km north of Akrounta, turn onto a forestry road (green sign) towards Apsiou and park after 200 m in an open space on the right (36 507121 E, 38 47800 N). About 50 m beyond the parking space is an outcrop in which two lava flows are separated by a thin unit of volcaniclastic rock containing fragments of serpentinite. This outcrop demonstrates that serpentinite must have been exposed at the sea floor at the same time as lava flows were accumulating near by. The tectonic movements that exposed the serpentinite must have taken place very early in the history of crustal accretion here.

Stop 4.6 Intrusions and shear zones in serpentinite of the Limassol Forest along the F128 between Akrounta and Arakapas This stop lies along the F128 between Akrounta and Arakapas, about 5 km north of Akrounta and 4 km north of the turnoff to stop 4.5. Coming from the south, park in an open space on the eastern side of the road after a long cutting on this side of the road (36 507918 E, 38 50194 N).

Start by walking about 300 m down hill to a straight section of roadcut in which very obvious mafic dykes cut the country rock, against which they are chilled. These dykes strike northeast–southwest, a common orientation for dykes in the Limassol Forest and one that is oblique to the trend of the Arakapas fault zone. The country rock is purple serpentinized harzburgite, with, in one place, a small gabbroic intrusion now transformed to white rodingite. The dykes have been metamorphosed to greenschist facies, but not transformed to rodingite, and the whole outcrop is cut by later brittle faults. This outcrop shows that the harzburgite was already serpentinized (and the gabbro rodingitized) before the intrusion of the dykes, and that the dykes were metamorphosed later.

Now walk back up hill to about 30 m short of the parking space. Here within the serpentinite there is a major shear zone that has been mapped for more than 2 km and runs east–west. The serpentinite is no longer massive and has a complex pattern of foliation that generally is vertical and striking east–west, but locally is deflected by lozenges of less-deformed serpentinite. Although not visible here, mafic dykes have been mapped as cutting this shear zone, so the shearing pre-dates the dyke intrusion.

A few tens of metres farther up hill, and below (west of) the road, are rounded brown outcrops. Walk through the bushes to examine them. These are part of an undeformed intrusion of wehrlite. In places it shows a clear cumulate texture, with rounded crystals of partly altered olivine surrounded by larger crystals of clinopyroxene. These features suggest that the intrusion must have taken place after the serpentinization of the harzburgite and after the generation of the large shear zone. Intrusion of the wehrlite presumably occurred at about the time of intrusion of the mafic dykes near by.

Stop 4.7 Low-angle extensional faulting within the Limassol Forest along the F128 between Akrounta and Arakapas This stop lies along a northwest–southeast-trending section of the F128, about 11 km north of Akrounta and 6 km from stop 4.6. The long roadcut of interest is on the southwestern side of the road, opposite a wooded valley below (36 508674 E, 38 52967 N). Park on the wide open space opposite the roadcut.

The roadcut is made predominantly of serpentinite that has a strongly developed foliation with an apparent dip to the southeast. In places along

Figure 4.6 Grey serpentinite capped by dark sheeted dykes (stop 4.7). These rocks are highly deformed and are separated by a near-horizontal extensional fault. The roadcut is about 4 m high.

the roadcut, slices of brown basaltic material are included within the serpentinite, and in others the roadcut is capped by a unit of mafic material separated from the serpentinite by a near-horizontal contact (Fig. 4.6). To see all of this may require a fair amount of walking back and forth. Several hundred metres northwestwards along the cutting, where the road swings sharply to the east, the rock is entirely mafic. A short distance farther on, this mafic material can be seen to be sheeted dykes.

All of these relationships suggest that the serpentinite and mafic material are separated by a major extensional fault, and the fabric in the serpentinite implies that the relative movement of the rocks above the fault was to the right when looking at the exposure. This is the most convenient roadside outcrop of a major extensional fault, although not the best (see stop 4.8). For those interested, there are excellent exposures of sheeted dykes farther along the road towards Arakapas.

Stop 4.8 Low-angle extensional faulting within the Limassol Forest near Akapnou   This stop lies to the east-southeast of the village of Akapnou, which lies just off the F114/F124 between Eptagoneia and Ora. At the roadsigns for Eptagoneia, Ora and Klonari, take the turning towards the village and turn left onto a narrow road after about 100 m, at a telephone box and fountain. Drive down this road, ignoring the turning to Klonari and Vikla, for 200 m to the village cemetery. Buses should park here. Walk or drive 800 m beyond the cemetery along a dirt road to a fork in the road; park cars here (36 518470 E, 38 55026 N) and walk southeast down the right-hand fork.

110

At first the roadside outcrops are brown sheeted dykes, but, after 200 m, outcrops abruptly become the pale-greenish colour of serpentinite where an extensional fault (poorly exposed here) is crossed (36 518660 E, 38 54899 N). Follow the road for 200 m to the bottom of the valley. Turn down a rough track to the right and cross the river (turnoff is at 36 518847 E, 38 54762 N). Twenty metres beyond the river, at the edge of an orchard, turn right onto a faint track and follow this up towards a ridge that runs down hill in front from left to right. Now follow the crest of this ridge up hill to the left (south) through outcrops of serpentinite. At the top of the ridge is an obvious left turn onto a terrace (36 518844 E, 38 54550 N), which should be followed towards the southeast. The back of the terrace is at first made of serpentinite, then of sheeted dykes (with some copper staining) and then, about 50 m farther on, of serpentinite again. At the end of this terrace continue to another higher terrace. The back of this higher terrace at first exposes serpentinite, but then, about 100 m from the far end of the first terrace there is a boundary between brown rock above and serpentinite below. This is a very fine exposure of one of the major extensional faults (36 518902 E, 38 54463 N).

Here, the base of the back of the terrace is serpentinite containing some discrete shear zones. About half-way up the outcrop is the low-angle fault, a sharp boundary separating about 50 cm of sheared serpentinite below from a bouldery breccia of mafic sheeted dykes. The apparent omission of major stratigraphical units, such as the gabbros, suggests that this is an extensional fault. The fabric in the sheared serpentinite below the fault shows that the relative movement of the rocks above the fault was towards the northwest, and within the fault is a lineation trending in the same direction.

It is worth spending some time exploring the area around this locality, both to see how the nature of the serpentinite changes with distance below the fault and also to take a broader look at the landscape. Along the side of the hill, the boundary between the brown-weathering sheeted dykes and the silvery serpentinite can be seen to run nearly horizontally for as far as the eye can see. About 1 km north is the Arakapas fault zone, behind which are hills of sheeted dykes belonging to the main body of the ophiolite. In the valley immediately below, the river runs through outcrops of massive brown serpentinite.

Stop 4.9 Gabbro in the Limassol Forest along the E109 between Kellaki and Parekklisia   About 3 km south of Kellaki and 5 km north of Parekklisia, the E109 runs down hill to the southeast, passing a very high and steep grey roadcut to the northeast, before bending towards the south. Park very carefully near this bend, especially as the road is dangerous because of fast traffic. Walk back up the road and climb up a faint trail onto a shelf half-way up the roadcut (36 514505 E, 38 50450 N).

The road is cut into layered gabbro, in which the layering dips steeply towards the northeast. The gabbro has been intruded by a swarm of later dykes that are chilled against the gabbro, and they strike northeast–south-west. The gabbro is interpreted as being part of a plutonic unit within a block of the hanging wall that rotated above one of the major extensional faults. The dykes are probably from the same phase of magmatism that produced the dykes at stop 4.6.

Chapter 5

The Mamonia terrane

Introduction

The geological foundations of southwest Cyprus are possibly the most complex on the island. Here there are a great variety of lithologies, interesting geological relationships and excellent exposures. The basement geology of this region is made up of rocks of the Cretaceous Troodos terrane and the Triassic–Cretaceous Mamonia terrane (Fig. 5.1), which were brought together by plate-tectonic processes in the Upper Cretaceous during closure of part of the Neotethyan Ocean. Erosional windows through the Maastrichtian and younger sedimentary cover rocks expose the suture zone between these two terranes. It is marked by outcrops of serpentinite and it forms an arcuate belt almost 30 km wide that runs east–west in the southwest part of the island and bends around to become north–south on the Akamas Peninsula (Fig. 5.1). These serpentinites are derived from the Troodos terrane and they form part of a series of displaced rock masses that resulted from the collision between the Mamonia margin and Troodos microplate. The partial subduction of the Mamonia margin beneath the Troodos microplate caused the build-up of a prism of accreted thrust slices at the subduction zone, in which Mamonia and Troodos rocks were tectonically mixed in an intricate fashion (see Fig. 1.3).

The presence of Mamonia-derived lithologies structurally interleaved with those from the Troodos terrane shows that collision between the two terranes was marked by extensive compressional tectonics. Where the thrust slices were dismembered, a chaotic mixture of lithologies resulted in a tectonic mélange. From a distance this mélange appears very colourful as it is made up of blocks of green serpentinite, greyish-brown volcanic rocks, white limestones, and grey-to-pink sandstones and cherts, all set in a matrix dominated by red mudstone. Reconstruction of the Mamonia stratigraphy and palaeoenvironment from this chaotic assemblage suggests that the Mamonia rocks represent a collapsed continental shelf and slope sequence that formed on the northern passive margin of Gondwana during the Upper Triassic formation of the Neotethys. The Jurassic to

113

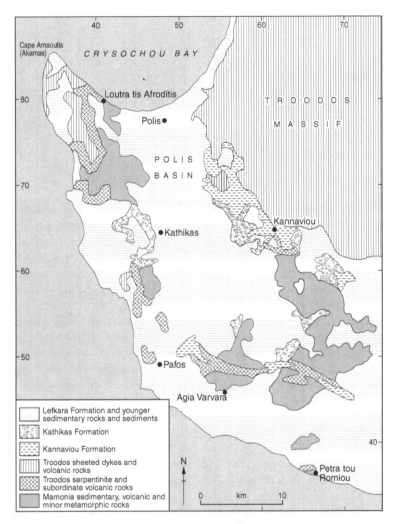

Figure 5.1 Simplified geological map of southwest Cyprus compiled from Lapierre (1971), Swarbrick (1980), Cyprus Geological Survey Department (1995) and Malpas & Xenophontos (1999). Note that not all maps agree on the areas covered by the Kathikas Formation and mélange of the Mamonia terrane. The numbers around the border are from the WGS84 UTM grid, where, for example, easting 50, northing 60 corresponds to the full grid reference 36 450000 E, 38 60000 N.

Lower Cretaceous period saw the development of this passive continental margin and its transition into a deeper-water sequence of clastic and pelagic sediments, which indicate overall regional subsidence. Associated with the deeper-water pelagic sedimentary rocks are reef-derived limestones and alkaline basaltic pillow lavas that are thought to represent

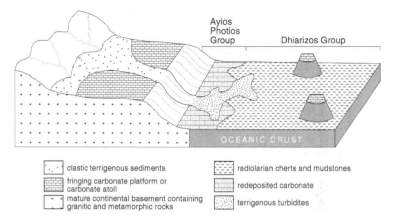

Figure 5.2 Schematic reconstruction of the upper Triassic passive margin represented by rock types of the Mamonia terrane (modified from Robertson 1990).

remnants of ocean islands, atolls or seamounts that formed on the older oceanic crust, which is itself preserved as occurrences of tholeiitic pillow lavas of Triassic age.

The reconstruction of the passive margin sequence permits division of the lithologies into two main groups: those deposited relatively close to the continent (the Ayios Photios Group) and those formed farther away (the Dhiarizos Group) (Fig. 5.2). The former consists dominantly of Triassic–Cretaceous clastic sedimentary rocks (sandstones, siltstones, mudstones, calcarenites and occasional limestones) and chert, which are preserved as well bedded sequences with sedimentary features suggesting deposition as immature turbidity-current deposits. Their composition indicates derivation from a mature continental landmass, which included granitic and metamorphic terranes that are not now represented in Cyprus. By contrast, the Dhiarizos Group is mostly made up of Triassic volcanic rocks, reef-derived limestones, radiolarian cherts and rare mudstones, all of which are thought to have formed some distance off shore from the continental margin. The volcanic rocks have compositions typical of within-plate tholeiitic and alkaline basalts that form seamounts in present-day ocean basins, and the limestones are envisaged to have formed fringing reefs around these volcanic pedestals.

The subduction of Neotethyan oceanic lithosphere beneath the Troodos microplate eventually resulted in the Mamonia continental margin arriving at the subduction zone. Here, because of its buoyancy, it effectively jammed the convergent system. However, some of the lithologies of the Dhiarizos Group were carried to some depth down the subduction zone, where they were metamorphosed to amphibolites, quartz–mica

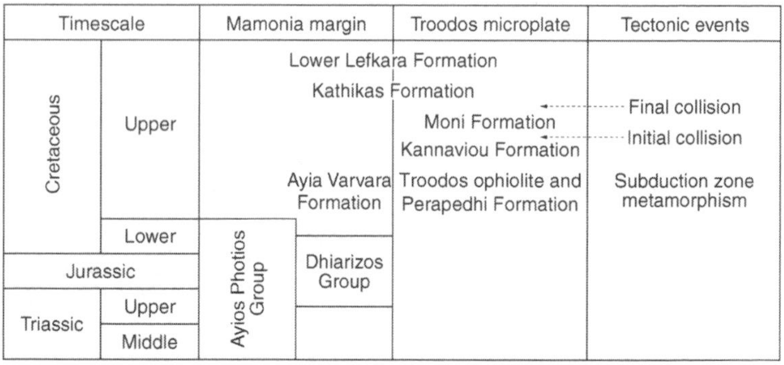

Timescale		Mamonia margin	Troodos microplate	Tectonic events
Cretaceous	Upper	Lower Lefkara Formation		
		Kathikas Formation		Final collision
			Moni Formation	Initial collision
			Kannaviou Formation	
		Ayia Varvara Formation	Troodos ophiolite and Perapedhi Formation	Subduction zone metamorphism
	Lower	Ayios Photios Group	Dhiarizos Group	
Jurassic				
Triassic	Upper			
	Middle			

Figure 5.3 Stratigraphy of the Mesozoic rocks of southwest Cyprus (modified from Malpas et al. 1993).

schists and marbles. These were later exhumed and incorporated into the suture zone during the intercalation of the Troodos and Mamonia terranes, and they are now known as the Ayia Varvara Formation (see Fig. 1.3).

The incorporation of serpentinites, gabbros and volcanic rocks of Troodos provenance into the suture zone suggests that the ophiolitic rocks involved in the collision must have resembled those in the Arakapas Valley and Limassol Forest of the Troodos ophiolite to the east. Here, serpentinites and gabbros were exposed on the sea floor as a result of high-level intrusion and extensive normal (extensional) faulting, resulting in east–west horst and graben structures. In fact, these graben acted as sites of deposition for a series of clastic sediments transported along their length by turbidity currents, which themselves have a unique composition. The sediments (the Kannaviou Formation) comprise ashes (now altered to bentonitic clay) and coarse volcanogenic sandstones derived from explosive eruptions that took place in an island-arc environment, for which there is little other evidence preserved in Cyprus, apart from in the Kyrenia terrane. These clays aided the formation of mass-wastage deposits or olistostromes (the Kathikas Formation), which formed because of the instability of the irregular submarine topography that was produced by the juxtaposition of the Troodos and Mamonia terranes, and which contain material from both terranes. The sequence of events recorded in the Mesozoic rocks of southwest Cyprus is summarized in Figure 5.3.

Mamonia rocks and contact relationships

The stops examine the wide variety of Mamonia and Troodos lithologies and their structural relationships as exposed in the erosional windows through the post-collision sedimentary cover. They are arranged geo-graphically, starting to the east-southeast of Pafos and ending to the north of this town, on the Akamas Peninsula. The Ayios Photios Group is examined at stops 5.2, 5.3 and 5.5, the Dhiarizos Group at stops 5.1 and 5.4, and the Ayia Varvara Formation at stops 5.7 and 5.9. Thrust faults and the tectonic mélange of the suture zone are specifically investigated at stops 5.6 and 5.8. Stops 5.2 to 5.5 can be combined into a single excursion up the Diarizos Valley, which takes 2–3 hours. For those wishing to explore aspects of the Mamonia terrane in greater depth, it is worth making an excursion up the Ezousa Valley north of the village of Agia Varvara, which lies off the E606 northeast of the A6 east of Pafos.

Stop 5.1 Sedimentary and volcanic rocks of the Dhiarizos Group at Petra tou Romiou   Petra tou Romiou lies along the B6 coastal road between Pissouri to the east and Kouklia to the west. There are three closely spaced localities within this stop, which mostly examine rocks of the Dhiarizos Group of the Mamonia terrane. The first locality provides an overview that will be recognizable as the scene on many postcards and in many guidebooks. To reach this locality, drive to a high point east of the tourist pavilion and park in a layby on the seaward side of the road (36 467021 E, 38 35586 N).

The spectacular view to the west reveals large blocks of pale limestone rising out of the inviting blue sea (Fig. 5.4). The largest block on the far promontory is Petra tou Romiou (the Rock of Romios). Thus, this rock is named not after Aphrodite, the goddess of love who is alleged to have ascended from the waves at the rock, but rather the Byzantine folk hero Romios, who used this and other large blocks and boulders to hurl at pirates. More importantly, from the viewpoint it is clear that the lime-stones are closely associated with dark rocks in the foreground, which are Upper Triassic volcaniclastic sandstones, tuffs, pillow lavas and lava breccias in a matrix of thin-bedded dark-red radiolarian cherts and mud-stones. This close link between the limestones and volcanic rocks is ubiq-uitous throughout the Mamonia terrane and lends support to the idea that the limestones represent what were once small carbonate reefs on volcanic islands and seamounts associated with oceanic crust that formed during the early stages of the opening of the Tethyan Ocean basin.

Before moving on, it is worth casting an eye over the slopes on the land-ward side of the road. These are dominated by chalks, marly chalks and marls of the Lefkara and Pakhna Formations, which are clearly unstable and failing in places because of movement concentrated within marl and within the clay-rich lithologies in the underlying Mamonia rocks and Kan-naviou Formation, upon which the carbonate rocks rest unconformably. This unconformity marks the end of significant regional tectonism in the

Petra tou Romiou

limestone

volcanic rocks

Figure 5.4 View west along the south coast to Petra tou Romiou (stop 5.1). The large white blocks of limestone in the sea are closely associated with dark volcanic rocks, and this association is typical of the Dhiarizos Group.

geological evolution of the island and the onset of a tectonically quiescent period in much of southern Cyprus.

The next two localities investigate the rocks forming both promontories to the west of the viewing area. To access the first, travel west down the road towards the tourist pavilion and pull off into the large layby on the seaward side of the road, just after crossing the first valley. From the eastern side of the parking area, follow the well trodden path that leads to the dark promontory. On the beach it is possible to examine closely a variety of dark-green volcanic rocks and blocks of white limestone. The most impressive exposure is that of pillow lavas and sheetflows (36 466528 E, 38 35835 N). The trachybasalt sheetflows form a wall-like exposure running into the sea and are loaded with white plagioclase crystals. They are sheared, faulted (note the slickensides) and very steeply dipping to the northwest. Sitting on top of these are impressive, steeply dipping, plagioclase- and pyroxene-phyric basaltic pillow lavas that have way-up indicators pointing towards the west. These indicators include radial cooling joints at the top of the pillows and highly amygdaloidal upper parts and rims of pillows. The latter are evidence that gases exsolved from the magma, but were unable to escape fully from the pillows because of rapid cooling and the pressure exerted by the overlying water body into which they were erupted. Extremely rapid quenching against sea water is evident from the ubiquitous glassy black pillow rims. Some pillows bifurcate into lobes, indicating a flow direction that is from the top of the exposure down towards the beach in the present outcrop.

Resting against the lavas are deceptive highly weathered volcaniclastic breccias that contain various clasts. It is easy to confuse them with Mamonia sedimentary rocks, especially mudstone, some of which has in fact slumped down from above. Opposite these volcaniclastic rocks are blocks of limestone that should be examined closely. Although most of these rocks are completely recrystallized and brecciated, there are several blocks of reef limestone that contain moderately preserved fossils as moulds, including corals, sponges, red algae, bryozoans, gastropods and bivalves. Most of these fossils are associated with stromatolitic material and suggest a Triassic age for the rock.

For the last locality, return to the main road and head to the foot of the hill, where there is parking associated with a souvenir shop and café. Take the pedestrian underpass to the beach to examine Petra tou Romiou. The most evocative feature of this large limestone block is the highly polished western face striated with slickensides, which demonstrate that blocks such as this were once jostled around in a tectonic mélange. Within the limestone are small volcanic clasts and a chilled pillow fragment, which have compositions of trachyandesite and trachyte.

Along the coast to the west are further outcrops of pillow lava, lime-stone and red deepwater sedimentary rocks. Faulted into these Upper Triassic rocks and exposed in the cliffs, are rocks, mainly serpentinite, of the Troodos ophiolite.

Stop 5.2 Sedimentary rocks of the Ayios Photios Group and Pakhna Formation along the F616 northeast of Nikokleia From the B6 between Kouklia to the east and Timi to the west, head northeast on the F616 towards Agios Nikolaos. At a distance of 5.6 km from the B6, and past the turnoff for Nikokleia, park on the right (east) on a flat area in front of a building and opposite a large roadcut through reddish Mamonia rocks on the inside of a sharp bend (36 463231 E, 38 43672 N). Be very aware of traffic when exam-ining the rocks along this bend.

The inside of the bend exposes highly disrupted sedimentary rocks of the Ayios Photios Group, which differ markedly on either side of a fault that strikes east–west. On the left (south) side the sequence is dominated by red mudstones intercalated with thin beds of reddish-grey fine-grain sandstones. On the right (north) side, medium- to thick-bedded turbiditic sandstones predominate, but the sequence also includes light-grey sand-stones and red mudstones. Examination of the grey sandstones, especially to the south of the fault, will reveal current marks produced by the impact of solid objects, superb load, flute and groove casts, and other way-up sedimentary features that show the sequence to be overturned. Similarly, to the north of the fault, the thick-bedded sandstones show very fine cross laminations, from which it can also be determined that these rocks are overturned. Some of the sandstones in the northern unit contain plant remains that have been carbonized and have reduced ferric iron to its fer-rous state, giving the rocks a greyish-green colour.

Farther north along the roadcut the sandstones become more calcare-ous and contain weathered-out clay intraclasts. The exposed sequence is topped by a massive light-grey calcarenitic sandstone and calcarenite that contain abundant quartz, feldspar, mica and ferromagnesian minerals (mainly amphibole), indicating a source dominated by granitic and region-ally metamorphosed rocks, which is not exposed in Cyprus. These calcare-ous rocks contain ooids, pisoliths, calcareous algae, echinoderm plates and benthic forams that suggest derivation from a carbonate platform in the Upper Triassic.

Walk south along the road for about 200 m until a roadcut on the west-ern side of the road is reached. Here, whitish chalks, marly chalks and calcarenites of the Pakhna Formation unconformably overlie reddish Mamonia mudstones (36 463092 E, 38 43599 N). The latter are poorly bed-ded and are deformed, and contain angular blocks of green lithologies, derived from the Troodos ophiolite, and some white limestone fragments. The Pakhna rocks are turbiditic and are bioturbated in the greyer layers, especially away from the contact with the Mamonia rocks. Pakhna calcarenites weather orange and contain occasional load casts and intraclasts. In some places the Pakhna rocks contain bioclastic layers, and in others they exhibit small-scale, probably syn-sedimentary, folds. The deformation recorded by the Mamonia mudstones pre-dates deposition of the Pakhna rocks.

The Diarizos Valley has had a rich human history. The abandoned village of Souskiou lies across the valley to the east and is set against a backdrop of cliffs, on top of which is a Chalcolithic settlement. Although thoroughly looted in the recent past, this site has produced a wealth of artefacts that are currently on view in the Pafos museum and the archaeological museum in Lefkosia. During the Ottoman occupation of Cyprus, the valley was appropriated by Turkish settlers because the river at that time flowed throughout the year and provided irrigation water for orchards and other agriculture. The names of many of the local villages (e.g. Fasoula, Agios Georgios and Agios Nikolaos) along and adjacent to the valley reflect the combined Greek and Turkish origins.

Stop 5.3 Cherts and mudstones of the Ayios Photios Group along the F616 northeast of Nikokleia Along the F616, 7.5 km from the B6 (1.9 km from stop 5.2), pull off on the right (east) side of the road, opposite an outcrop of well bedded sedimentary rocks (36 464270 E, 38 45029N).

In the roadcut are red to light-grey thinly bedded deepwater sedimentary rocks consisting mostly of cherts and minor mudstones (Fig. 5.5a). In the grey cherts, radiolarians can be seen through a hand lens, especially on wetted surfaces. These cherts were deposited in deep water below the carbonate compensation depth sometime in the Jurassic, which suggests that they are part of the Ayios Photios Group. Towards the top of the roadcut there are paler and slightly thicker beds of pelagic limestone that contain diagenetic replacement chert. Their existence at first seems problematic, because they are associated with deepwater sedimentary rocks. However, closer examination reveals that the limestones are calcareous turbidites, which implies that they must have been deposited rapidly below the carbonate compensation depth from high-density currents and then covered before carbonate dissolution occurred. In the distance, on the opposite side of the road, red sandstones are quarried for road metal to upgrade the F616.

Stop 5.4 Basaltic lavas of the Dhiarizos Group near Fasoula Following the F616 to just north of Fasoula and a small café, there is a prominent ridge

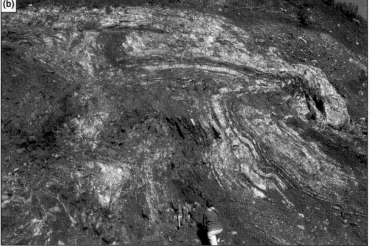

Figure 5.5 Well bedded cherts and mudstones of the Ayios Photios Group: **(a)** undeformed (stop 5.3); **(b)** complexly folded and sheared (stop 5.5). The height of (a) is about two thirds of (b).

on the left-hand (western) side of the road, just beyond where a paved and an unpaved road join the F616 from the west. Park where these two roads join the F616, which is 10.1 km from the B6 and 2.6 km from stop 5.3. Walk up the unpaved road for about 30 m, to where a concrete irrigation channel meets the track, and climb up to the path above the channel to view the rocks (36 465899 E, 38 46673 N). The remarkable preservation of the rare features at this location means that there must be no hammering or removal of rock from the area.

At this locality, well formed dark greenish-grey olivine-phyric pillow lavas are exposed, but the olivine has been replaced by a brownish mixture of iron oxide and carbonate minerals. The pillows have glassy margins, but the glass has mostly been altered to palagonite. Underneath these margins, and best seen where they have been weathered away, the pillows exhibit a superb variolitic texture defined by small spherules that give exposed surfaces a pockmarked appearance. This texture is indicative of rapid quenching of lava, which caused supercooling of the liquid and localized growth of some crystals prior to the onset of rapid crystallization of the remaining liquid. Geochemically these pillow lavas are olivine tholeiites typical of those found on intra-oceanic volcanic islands, and they are part of the Dhiarizos Group.

Within some pillows there are obvious arcuate bands composed of the white zeolite, laumontite. These bands are confined to the uppermost parts of the pillows and they follow the shape of the top surface of pillows. It is thought that the semi-circular cavities in which the zeolite formed were produced when lava partially drained from feeder tubes in which the flow of lava may have been pulsating. The successive bands of zeolite may have been produced during periodic streaming of hydrothermal fluids through the cavities. Laumontite also occurs as beautiful radiating crystals in between pillows.

Stop 5.5 Deformed cherts and mudstones of the Ayios Photios Group at Mamonia At a distance of 12 km along the F616 from the turnoff from the B6 (1.9 km from stop 5.4), there is a coffee shop (the Old Village Mill) just beyond the last house at the northern end of Mamonia village. Here there is parking (36 467044 E, 38 47693 N) and a spectacular outcrop of folded Mamonia sedimentary rocks in a 12 m-high roadcut that is topped by colluvium and old river gravels (Fig. 5.5b).

The sedimentary rocks are part of the Ayios Photios Group. They comprise thinly bedded yellow cherts and red mudstones, with at least one major intercalation of thick-bedded purple-to-red coarser-grain calcareous sandstone that exhibits grading. There are patches of replacement chert in the coarse sandstone, as well as primary angular limestone and red chert clasts, suggesting that this is an immature sedimentary rock, probably a turbidite. Chert nodules are also present. The rocks at this stop are similar to those at stop 5.3 and indicate an origin in deep water, with the carbonate-rich lithologies surviving below the carbonate compensation depth because of rapid deposition and burial.

The whole sequence is complexly folded; the folds are asymmetric with the large flat limbs sheared in places and the shorter limbs more intricately folded. The main closure of the folds is to the east-northeast. The deformation is tectonic in origin and associated with the collision of the Mamonia margin with the Troodos microplate.

Stop 5.6 The suture zone, tectonic mélange and geological hazards between Nea Choletria and Nata This stop requires half a day to investigate six localities distributed over a distance of about 5 km between Nea Choletria and Nata (Fig. 5.6). Together, the localities provide an overview of the structural complexity and lithological variability of the Mamonia–Troodos suture zone and the tectonic mélange of the Mamonia terrane. They also illustrate some of the geological hazards typical of southwest Cyprus. Start in Nea Choletria, which lies along the F617 between the F616 and B6 to the southwest and Stavrokonnou to the northeast. On the northern side of the village, park on a large flat area on the western side of the road, opposite a coffee shop and other establishments above the road (36 463290 E, 38 47000 N; Fig. 5.6, point a).

The scene west across the valley of the Xeros Potamos is superb, especially in the morning. The full view of both sides of the valley shows that the colourful and chaotic Mamonia mélange is exposed in an erosional window through the overlying white sedimentary rocks that belong mainly to the Lefkara and Pakhna Formations. On the distant horizon these cover rocks are exposed to the south and north of darker rocks, which include Mamonia sedimentary and volcanic rocks, Troodos serpentinites and volcanic rocks, and Kannaviou clays. In favourable light it is possible to see that some of these dark rocks form a stack of thrust slices defined by the inclined stepped appearance of the distant hilltop.

Now continue northwards on the F617, first up hill and then down hill to the northeast. Along this downhill section, just over 1 km from the last locality, there is parking on the southeastern side of the main road where it is met by a small road (36 463820 E, 38 47790 N; Fig. 5.6, point b). Begin with an examination of the roadcuts along the main road to both sides of the parking area. Their geology is extremely complicated and chaotic because several lithologies have been brought together by shearing and a thrust fault. The thrust contact can be seen in the cutting immediately above the parking area, where highly sheared and chloritized Dhiarizos lavas and small slivers of Ayia Varvara amphibolite overlie reddish mélange. The structure, the abundance of clay-rich lithologies, and the steepness of the roadcuts make these slopes highly unstable. In an attempt to reduce the risk they pose, the highest slopes have been terraced. Further evidence of instability can be seen on the crest of the hill to the southwest. Here, repeated erosion, slumping and flow of the clay and silt matrix occurs during and after heavy rainfall, often leading to the displacement of large competent blocks of rock. Concrete drainage ditches have been

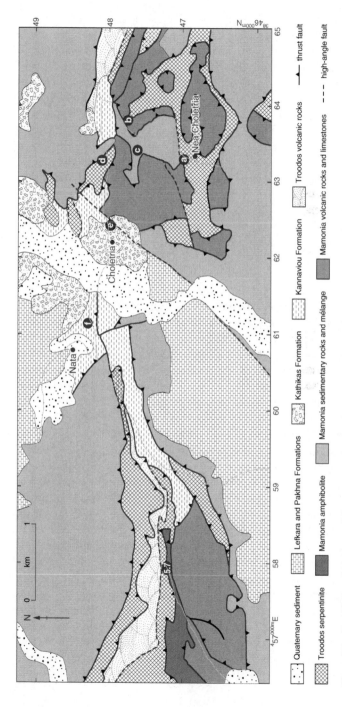

Figure 5.6 Simplified geological map of the area around and to the west of Choletria (after Malpas & Xenophontos 1999). Points a–f correspond to the six localities investigated within stop 5.6. The numbers around the border are from the WGS84 UTM grid.

Quaternary sediment

Troodos serpentinite

Lefkara and Pakhna Formations

Kathikas Formation

Mamonia amphibolite

Mamonia sedimentary rocks and mélange

Kannaviou Formation

Troodos volcanic rocks

Mamonia volcanic rocks and limestones

▲ thrust fault

▲ high-angle fault

constructed to prevent slopewash, mudflows and rockfall reaching the road, but the slopes continue to fail and they pose a threat to road traffic.

Near the crest of the hill there is a narrow dirt track leading off to the west (36463607 E, 3847677 N), which should be followed until the first field on the left (south) is reached. From here, head southwest to a knoll of volcanic rocks a short distance off the track (36463393 E, 3847634 N; Fig. 5.6, point c). The rocks are sanidine-bearing trachytes and trachy-basalts that crop out as dark greenish-grey sheetflows. The elongate, rectangular and equant crystals of off-white sanidine in places exhibit simple twinning that may be seen with the naked eye or through a hand lens.

From the knoll, walk north across a field to a prominent ridge exposing dark pillow lavas to the west and light-grey pillow lavas to the east (36463318 E, 3847686 N). The light-grey pillow lavas are plagioclase-phyric alkali basalts in which phenocrysts are up to several centimetres across. Some pillows have centres enriched in plagioclase phenocrysts, and margins depleted in these crystals. This feature indicates that the phenocrysts must have formed in a magma chamber prior to their transport through lava tubes during a volcanic eruption. Inter-pillow cavities have been infilled by white and red limestone made predominantly of very fine-grain calcite. These lavas underlie, but also interfinger with, darker fine- to medium-crystalline alkali basalts. The contact is clearly exposed and worthy of closer examination. Both types of lavas are amygdaloidal and in some cases concentrations of amygdales towards the tops of pillows indicate that the lavas are the right way up. Once again, the close association of lavas and limestone in the Dhiarizos Group is evident from a large block of limestone immediately to the west of the lava outcrop.

Return to the parking area and continue a short distance down hill on the F617, before taking the first turning to the west for Agios Panteleimon, Choletria and Nata. Travel 850 m down this road and park at the side of the road, some 20 m south of a prominent outcrop of dark-green serpentinite (36463297 E, 3848127 N; Fig. 5.6, point d). Although the road is narrow, it should be safe to park because there is normally very little traffic. Walk over to the serpentinite, from where there is an excellent view of the area westwards to Nata across the Xeros Potamos. It is clear that this is only one of several serpentinite outcrops forming a discontinuous ridge that is the southern boundary of a broad east–west-trending valley. The northern limit of the valley is seen in the distance, marked by fewer serpentinite outcrops. The low ground within the valley is occupied by a sliver of Troodos lavas covered by bentonitic clays of the Kannaviou Formation and by red debrites of the Kathikas Formation. The serpentinites delineate the two bounding faults on either side of the Troodos volcanic rocks: Mamonia sedimentary rocks to the north and Mamonia volcanic rocks to the south.

The outcrop of serpentinite is highly brecciated and sheared, with prominent interconnecting shears that trend almost east–west and dip steeply to the south. A closer examination of the serpentinite clasts reveals millimetre-size crystals of altered orthopyroxene in a green fine-grain matrix, suggesting that the protolith was harzburgite. Along its southern margin facing the road, the serpentinite is in contact with Kannaviou bentonitic clays that are partly covered by red clays washed down from the Mamonia rocks on the other side of the road to the south. Along this contact and for a short distance into the outcrop, the serpentinite is highly silicified, and quartz also infills fractures. Such modification of the serpentinite is found wherever it is overlain by bentonitic clays, raising the possibility that the clays are the source of the silica. This is highly likely, as the clays are derived by the alteration of silica-rich volcanic ashes, a process that liberates free silica.

The brecciated nature of the serpentinite outcrop suggests that it is representative of a talus deposit formed by mass wastage off a scarp of serpentinite. More significant, however, is the implication that serpentinite had been exposed on the sea floor to be eroded and form the talus pile upon which the Kannaviou clays were deposited, albeit locally.

Return to the road and travel down hill for just over 1 km in order to park along the road near two concrete water tanks (36 462415 E, 38 48019 N; Fig. 5.6, point e) at the eastern end of Choletria. From here, explore the village, keeping an eye out for snakes. Choletria consists of an assortment of mainly derelict buildings, the most prominent of which is the roofless church (36 462123 E, 38 48040 N), which stand on the Kathikas Formation of reddish clays that are interbedded with shales, sandstones and siltstones, and occasionally contain blocks of limestone and serpentinite. The high clay content of the foundation material makes the area susceptible to mass movements induced by earthquakes or heavy rain.

Choletria was damaged by the major earthquake of 10 September 1953, but not severely enough to warrant relocation. However, during the 1960s, further mass movement occurred during and after periods of heavy rainfall. Although movement was limited by the stabilizing effect of more competent layers in the clay, buildings did suffer structural damage, some even collapsed, and water escaping from sandstone emerged through the floors of many houses. Evidence of structural damage is extensive and it varies from building to building according to the type of construction material used and the underlying geology. Walls made of adobe (straw and mud) bricks are bowed and cracked owing to their greater plasticity relative to the brittle failure exhibited by walls constructed from stone, fired bricks and breeze blocks.

Unfavourable foundation geology and extensive building damage prompted relocation of the village onto Troodos serpentinites and Dhiarizos lavas at the top of the hill to the southeast (Fig. 5.6, point a). Despite the risk of future mass movement in the area, new houses are being built in Choletria. It is worth considering the potential stability of the ground upon which they are built. For example, take note of the new house on top of the prominent and stable block of serpentinite to the northwest.

From Choletria, head west across the Xeros Potamos, bearing in mind that it may not be fordable after heavy rain, and follow the road up hill to a large exposure of greenish-grey bentonite of the Kannaviou Formation that lies just below the first modern house on the eastern outskirts of Nata. It is possible to park on the track leading up to this house (36 461124 E, 38 48294 N; Fig. 5.6, point f). The view to the east is superb and it puts in context many of the localities examined so far (Fig. 5.7).

Bentonite is exposed directly up slope from the parking area, where it appears to overlie harzburgite, and it contains fragments of chalk talus washed down from higher up the slope. The principal component of the bentonite is the montmorillonite group of clay minerals, but colloidal silica is also present. The bentonite is soft, plastic and porous, and in wet weather it absorbs water, expands and flows down slope, as is evident from the deep gullying of the exposure. Conversely, during hot dry weather the clay becomes highly desiccated and it shrinks and cracks. It is, therefore, highly unsuitable as a foundation material, but it has wide application

Figure 5.7 The eastern slope of the valley of the Xeros Potamos as seen from Nata (point f of stop 5.6). Nea Choletria sits on a foundation of Troodos serpentinites and Dhiarizos lavas, whereas the remains of Choletria (centre of view) rest on rocks of the Kathikas Formation that lie at the base of a steep slope dominated by Dhiarizos lavas.

where large quantities of water must be absorbed, such as in the cut-off trenches of dams. The high absorptive capacity of the clay means that much of what is extracted locally and elsewhere in Cyprus is dried, powdered and bagged, and shipped to Europe for use as cat litter. A small-scale bentonite extraction, processing and bagging operation is situated on the western side of the Xeros Potamos, just north of where it is crossed by the road from Choletria.

Continue into Nata for refreshments and further consideration of problems of land instability. Parts of Nata are built on bentonite and chalk talus, both of which are unstable and have led to movement in building foundations, prompting many residents to abandon their homes. Nevertheless, some of these buildings have been renovated for use or sale, but Choletria should serve as a reminder of what could happen in the future.

Stop 5.7 Metamorphic rocks along the E606 northeast of Agia Varvara This stop examines a roadcut along the E606, 3.8 km northeast of the turnoff for Agia Varvara and 2.8 km south-southwest of the turnoff for Nata (see Fig. 5.6). There is parking where a dirt road joins the E606 from the southeast (36 457914 E, 38 47367 N). If coming from the direction of Nata, note the failing section of the E606 where it passes over red Mamonia and grey Kannaviou clays at the head of a V-shape wedge of slope.

The roadcut exposes part of a thin east–west-trending sliver of metamorphic rocks of the Ayia Varvara Formation, which has thrust-faulted contacts with Troodos serpentinites to the north and with Dhiarizos lavas to the south (see Fig. 5.6). The outcrop consists of a series of interbedded metasedimentary (mostly quartzite) and metavolcanic (amphibolite) rocks, which have a marked foliation that runs east–west and dips 50° north. The amphibolites consist predominantly of epidote–hornblende schists, which preserve rare small isoclinal folds.

Although the rocks are highly deformed, the intercalation of what were originally sedimentary and igneous rocks is not considered to have been produced tectonically, but is more reminiscent of the close association of volcanic and sedimentary rocks found elsewhere in the Mamonia terrane as the Dhiarizos Group. On the basis of this association, the metamorphic rocks are considered to represent silica-rich sedimentary rocks and altered volcanic rocks of the Mamonia terrane. These original rocks were subjected to high temperatures and pressures associated with the formation of an accretionary prism in a subduction-zone environment (see Fig. 1.3).

Stop 5.8 The suture zone, rocks of the Troodos and Mamonia terranes and mass movement along the Mavrokolympos Valley From the E701 immediately adjacent to the coast between Pafos and Coral Bay, and north of the village of Kissonerga, take the signposted road inland towards the Mavrokolympos dam. Park before the first roadside exposure, which occurs where the land begins to rise steeply (36 444431 E, 38 56482 N), and proceed on foot along the track for about 2 km, making observations along the way.

The Mavrokolympos Valley has been cut down through carbonate rocks of the Lefkara and Pakhna Formations to produce a window into a thrust sequence of rocks belonging to both the Troodos and Mamonia terranes. Broadly, in order of appearance from the first exposure, the dirt track leading up the northern side of the valley passes grey and brown serpentinite and sheeted dykes of the Troodos terrane, red Mamonia mudstones and siltstones of the Ayios Photios Group that exhibit chevron folding, Troodos serpentinites overlying light greyish-green Troodos lavas immediately west of the dam, Troodos serpentinites and Dhiarizos lavas and limestone at the dam, and extensive red cherts and mudstones and grey sandstones of the Ayios Photios Group beyond the dam to the east. The most impressive exposures are those near the northern abutment of the dam, where highly sheared serpentinites exhibit thrust-faulted contacts with deformed reddish-brown Mamonia sedimentary rocks and, underneath, Troodos pillow lavas. A host of structural features can be seen here, including boudins and deformation fabrics (Fig. 5.8).

Clearly, the geology along the track and in the surrounding area is structurally complex. It resulted from two contractional fault systems in the area, which both display similar stratigraphical stacking sequences.

Figure 5.8 Highly sheared and altered slivers and boudins of Troodos and Mamonia rocks in a thrust zone (stop 5.8).

They consist of a basal zone of sheared serpentinites overlain by thrust sheets of rocks of the Dhiarizos Group, which are in turn overlain by repeatedly overlapping thrust sheets of the Ayios Photios Group. Structural relationships show that the basal portions of the Mamonia thrust systems truncate and, therefore, postdate the lithological units and extensional structures of the underlying Troodos terrane. Along the section walked, most of the thrusts are generally west directed and define a duplex structure. On a regional scale, these west-directed (and associated south-directed) faults postdate the earlier north- and east-directed ones that initially placed Mamonia rocks over Troodos rocks during the collision of these two terranes.

The Mavrokolympos earth dam was constructed between 1964 and 1966 without prior geological site investigation. The northern abutment is keyed into a large mass of highly sheared and thrust-faulted serpentinite and blocks of Dhiarizos lava and reddish-stained limestone, whereas the southern side of the dam abuts red Ayios Photios mudstones and siltstones that occasionally contain competent blocks. The base and sides of most of the reservoir are composed of these red clay-rich Mamonia sedimentary rocks, which are prone to movement, as is evident in the arcuate slump scars and hummocky topography of the valley sides. Much of this mass movement occurred shortly after completion of the dam and was forceful enough to transport and damage concrete tanks that were built on the valley sides during construction of the dam. An example of one of these can be seen on the northern shore of the reservoir. The mudstones immediately up stream of the top of the spillway have been smoothed and a surface drainage channel constructed in order to prevent further slope failure. Other stabilization measures include the flattening of the slope along the northern bank of the reservoir and the placement of a rock-toe embankment at the base of this slope.

Stop 5.9 Deformed and metamorphosed volcanic and sedimentary rocks at Loutra tis Afroditis Loutra tis Afroditis (Baths of Aphrodite) is clearly signposted and lies at the western end of the E713 west of Prodromi. From the parking area, take a track on the eastern side of the tourist pavilion and restaurant that leads down to a rocky and pebbly beach, and then walk southeast to an impassable headland (36 440632 E, 38 79393 N). The aim from here is to examine the rocks and structures along the coast for several hundred metres to the northwest. Exercise caution as the rocks are slippery and it is possible to be cut off by the sea, especially in windy weather.

The traverse mainly examines rocks of the Mamonia terrane, but at the northwestern end Troodos serpentinites are exposed in the cliffs. The impassable headland comprises a dark reddish-brown to grey volcaniclastic breccia containing angular and sub-rounded clasts of basaltic and trachytic lava, some of which resemble fragments of pillows. These lavas are the equivalent of those that are better preserved at stops 5.1, 5.4 and 5.6. The next prominent outcrop exposes dark reddish-brown cherts and mudstones containing zones of black manganese oxide. The rocks in these two outcrops belong to the Dhiarizos Group.

To the northwest of the point where the track meets the beach, there are outcrops containing deformed mudstone, chert, quartzite, massive recrystallized and banded limestone, and marble. These rocks have been affected by several phases of deformation that resulted in folding and faulting.

After passing a point where steps come down to the beach, there is a promontory (36 440473 E, 38 79654 N) of dark amphibolite that records at least three phases of deformation; for example, close inspection will reveal that isoclinal folds have been refolded into kink folds. The amphibolite contains the minerals amphibole, epidote and plagioclase, and is the metamorphic equivalent of alkaline within-plate volcanic rock. The pinkish island off shore from the promontory is composed of low-grade metasedimentary rocks, but farther northwest, around the promontory, there are deformed higher-grade rocks dominated by quartz–mica schists and quartzites (Fig. 5.9). The rocks forming the promontory are part of the Ayia Varvara Formation, and were produced during metamorphism of volcanic and sedimentary rocks of the Dhiarizos Group as they were subducted during collision of the Mamonia margin with the Troodos

Figure 5.9 Highly contorted pale quartzites and darker quartz–mica schists of the Ayia Varvara Formation (stop 5.9). The thickest quartzite bands are 2 cm wide.

microplate. Farther along the coast there are bodies of the Ayia Varvara Formation associated with Troodos serpentinite, the first occurrence of which is marked by a steep fault zone.

If time permits, it is worth returning to the car-park and walking the Adonis and Aphrodite guided nature trails (each one is about 7.5 km long). One of the early stops along the trails (they are the same trail to begin with and then they split) is the Baths of Aphrodite, which is a pool and grotto. Legend has it that this is where Aphrodite bathed and also where she met her lover Adonis for the first time when he stopped to quench his thirst after hunting in the forest of the Akamas. The legend also states that everyone who drinks from the spring feels younger and more loving. In fact those who drink the water are likely to fall ill, because it is not potable, so do not drink it. The pool exists because the water table is perched at a high elevation within permeable carbonate rocks of the Pakhna Formation. These rocks are underlain by less permeable serpentinite, and water issues from springs in Pakhna rocks above the serpentinite.

For those interested in archaeology and coastal scenery there are some superb ancient quarries (36 437630 E, 38 81570 N) that can be visited from Loutra tis Afroditis by following the E4 coastal track for 3.5 km northwest towards Fontana Amorosa. The quarries lie in a sheltered bay at the edge of a relatively flat coastal plain, which is a Holocene marine terrace composed of calcarenite. It is this rock that was quarried, probably by the Romans, and used as building stone in Nea Pafos (see stop 7.16) or the ancient city of Marion–Arsinoe where Polis now stands. The calcarenite makes a reasonable building stone because it is relatively free of obvious bedding and joints, but its quality is compromised by its porosity and susceptibility to chemical weathering, as can be seen in outcrop.

Chapter 6

The circum-Troodos sedimentary succession

Introduction

The circum-Troodos sedimentary succession records over 80 million years of geological history, during which time the Tethys Ocean closed to leave the Mediterranean Sea, and Cyprus as an emergent island (Table 6.1). The sedimentary cover developed somewhat differently in areas around the Troodos massif. To its south, west and east, Upper Cretaceous to Miocene pelagic chalks, marls and calcarenites dominate the scenery, whereas to the north of the massif, clastic sedimentary rocks of Pliocene and Quaternary age occur in great abundance (see Fig. 1.2).

Deposition of the sedimentary cover continues to take place in a mainly compressional tectonic regime that has dominated since the collision of the Troodos and Mamonia terranes in the Upper Cretaceous (Maastrichtian), when the Moni and Kathikas Formations developed. Since that time, the continuing effects of plate convergence have resulted in a 90° anti-clockwise rotation of the Troodos microplate, pulses of compression and extension, the development of local basins and isolated highs, and the rapid uplift of the Troodos massif, which peaked in the early to mid-Quaternary (Pleistocene). The uplift of Cyprus was mostly domal and centred around Mount Olympos, and it resulted in various parts of the island being elevated and flexed by different amounts, with some showing direct vertical uplift, others tilting both up and down, and some showing little or no apparent elevation at all. The last category is exemplified all along the southern fringe of the island, where rocks that formed in shallow water in the Upper Miocene still lie in shallow water today. This is in stark contrast to the elevated Quaternary marine terraces in parts of south and west Cyprus and on the north and south sides of the Kyrenia Mountains. These suggest that the western side of Cyprus has been uplifted more than the eastern side, which even appears to have dropped in places, as exemplified by the partial submergence of the ancient city of Salamis.

Table 6.1 The circum-Troodos sedimentary succession stratigraphy (based on Cyprus Geological Survey Department 1995).

Age	Formation	Sedimentary rocks and sediments	Events
Holocene	Alluvium–colluvium		Substantial pulsed uplift and erosion of the Troodos massif mainly in the Pleistocene
Pleistocene	Terrace deposits	Calcarenites, sandstones, conglomerates, marls, clays, silts, sands and gravels	
	Fanglomerate		
	Apalos–Athalassa		
	Kakkaristra		
Pliocene	Nicosia		Marine transgression due to reconnection of the Mediterranean basin with the Atlantic Ocean
Miocene	Kalavasos	Gypsum, chalks and marls	Messinian salinity crisis arising from the isolation of the Mediterranean Sea from the Atlantic Ocean
	Pakhna	Hemipelagic chalks and marls; calcarenites, conglomerates and reef and bioclastic limestones	Uplift and initial emergence of the Troodos ophiolite as a consequence of northward subduction along the Cyprean arc. Development of the Yerasa fold and thrust belt
Palaeocene to Oligocene	Lefkara	Pelagic chalks and marls with chert bands and nodules in places	Deep-marine sedimentation during a tectonically quiet period
	Kathikas	Debrites with a clay-rich matrix	Post-collision submarine mass movement
Maastrichtian	Moni	Undeformed and deformed (mélange) bentonite-rich sedimentary rocks	Collision between the Troodos microplate and Mamonia margin and early rotation of the Troodos microplate
	Kannaviou	Volcaniclastic sandstones and bentonitic clays	Calc-alkaline volcanism outside the modern boundaries of Cyprus
Campanian	Perapedhi	Umbers, shales and mudstones	Hydrothermal and deepwater sedimentation

The Lefkara Formation records some 50 million years of tectonic quiescence, but the end of this stability is marked in the overlying Pakhna Formation by lithologies and structures that record rapid differential uplift of the Troodos massif in the Miocene, during which time several discrete basins began to form and evolve (Fig. 6.1). During the Middle Miocene, compression and uplift resulted in southwestward-directed thrusting of ophiolitic rocks of the Limassol Forest and associated deformation of adjacent Lefkara and Pakhna rocks along the Yerasa fold and thrust belt. By the late Middle Miocene, areas of significant uplift were centred on Mount Olympos, the Limassol Forest and the Akamas Peninsula, and these were eroding and yielding significant quantities of clastic sediment. Much of the Troodos massif emerged as an island in the Middle to Upper Miocene, about 7 million years ago. Since then, the central ridge of the Troodos massif has risen at least 4 km, while erosion has compensated for this and cut deeply into the emergent mountains. Nonetheless, in many places the topography still reaches 1–2 km above sea level.

The basins around the emerging Troodos massif permitted extensive crystallization of evaporite minerals during the uppermost Miocene (Messinian) salinity crisis, a time when the Mediterranean Sea became isolated from the Atlantic Ocean. In central Cyprus, the Mesaoria basin (the largest basin) first subsided at this time and over 1 km of marine sediment subsequently accumulated throughout the Pliocene, when the Mediterranean basin and Atlantic Ocean were reconnected. The Mesaoria basin began to

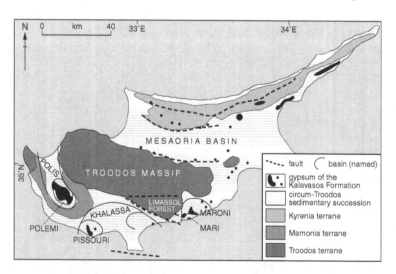

Figure 6.1 Major structural and topographical elements in Cyprus and occurrences of gypsum of the Kalavasos Formation. Adapted from Cyprus Geological Survey Department (1995) and Robertson et al. (1991, 1995).

emerge in the Quaternary, when the sedimentary fill became fluvial rather than marine. In western Cyprus, the Polis basin began to subside at about the same time as the Mesaoria basin, but it is now uplifted and dissected by rivers that feed sediment along its axis to parts that still lie below sea level. The Polis basin is an outstanding example of an extensional (graben) structure within an overall compressional regime and it is bounded by the Akamas Peninsula to the west and the Troodos massif to the east (Figs 6.1, 6.2). Southwest of Lemesos, the Akrotiri salt lake lies close to sea level and covers another sedimentary basin that is just beginning to emerge from the marine environment.

The Perapedhi Formation

The circum-Troodos sedimentary succession has at its base the Upper Cretaceous (Campanian) Perapedhi Formation resting on top of the igneous rocks of the Troodos ophiolite. The formation comprises distinctive brown umbers that locally grade upwards into deepwater radiolarian mudstones and marly sedimentary rocks. The umbers are a derivative of hydrothermal venting on the sea floor and are, therefore, an integral component of the Troodos ophiolite. The formation is best developed along the northern margin of the Troodos ophiolite, where umber may be clearly seen filling in depressions in the lava surface. Along the southern margin, although much less well developed, the Perapedhi Formation does in places grade upwards into the Kannaviou Formation. The Perapedhi Formation may be examined at stops 3.14, 3.15, 3.16, 3.18, 4.4 and 7.3.

Kannaviou, Moni and Kathikas Formations

The Kannaviou and Kathikas Formations crop out in southwest Cyprus (see Fig. 5.1), but the Moni Formation is restricted to an area north and east of Lemesos. As clay is a major component of all of these formations, they are inherently unstable and they slump and flow down slope to produce an uneven terrain characterized by rotational slope failures, hummocks and lobes.

The grey Upper Cretaceous (Campanian to Maastrichtian) Kannaviou Formation is made up of bentonitic clays, derived from weathering of fine-grain volcanic ash, and of volcanogenic sandstones, both of which developed in patches on top of the ophiolite about 20 million years after it had formed. Where it is *in situ*, the Kannaviou Formation rests only on rocks of the Troodos ophiolite, including those of the Perapedhi Formation, and

Figure 6.2   The northern end of the Polis graben, looking northeast from the E709 north of Kathikas. The Troodos ophiolite makes up the highest hills in the far distance.

pre-dates the collision with the Mamonia margin (see Fig. 5.3). The Kannaviou Formation should, therefore, be considered to be a component of the Troodos terrane. However, it is not part of the ophiolite, as the volcanoes from which the Kannaviou sediments were derived erupted explosively, were of calc-alkaline composition, and were probably located somewhere outside the current boundaries of Cyprus. After deposition, the Kannaviou sediments were involved in the formation of the chaotic and coloured mélanges that are components of the Moni and Kathikas Formations.

The collision of the Mamonia margin with the Troodos microplate caused anti-clockwise rotation of the Troodos microplate, tectonic mixing of Mamonia rocks and fragments from the Arakapas transform fault, and concomitant uplift and exhumation of the accretionary prism that had formed at the subduction zone. This process resulted in a series of deformed chaotic sedimentary assemblages, essentially large-scale mass-wastage deposits or huge debris flows, that covered and effectively sealed the suture between the Troodos and Mamonia terranes (see Fig. 5.3). These are Upper Cretaceous (Maastrichtian) in age and are known as the Moni Formation where observed south and southwest of the Limassol Forest and as the Kathikas Formation in the Pafos district in the southwest of the island. The generation of these mélanges, which may have both tectonic and sedimentary origins, was promoted by extensive thrust and strike-slip faulting associated with plate collision and rotation. In addition, instability and sliding were aided by the clay-rich matrix of the older Kannaviou Formation, which was mixed in from the Troodos microplate during plate collision. The debris flows are depositionally interbedded with pelagic carbonates of Upper Maastrichtian age that were deposited during hiatuses between mass wasting and associated deposition.

The Moni Formation appears to lie stratigraphically between the older Kannaviou and younger Kathikas Formations, but it is a somewhat enigmatic unit and some workers disagree with its classification as a formation in its own right. It occupies a trough along the southern and southwestern extent of the Limassol Forest of the Troodos ophiolite. It is characterized by grey bentonitic clays that may be undisturbed or deformed, the latter forming a mélange. Undisturbed portions of the formation appear to have normal depositional contacts with the underlying ophiolite and Perapedhi Formation and the overlying Lefkara marls and chalks. In many ways these parts are equivalent to the Kannaviou Formation in terms of lithology, age and depositional setting, and these similarities have led some researchers to suggest that the Moni Formation is in fact part of the Kannaviou Formation.

The youngest portion of the Moni Formation is a mélange, which some workers also assign to the Kannaviou Formation, but others state that it forms a separate entity that postdates the Kannaviou Formation and is the first record of the impending collision of the Mamonia margin with the Troodos microplate. The mélange comprises blocks and sheets of a wide variety of lithologies that may exceed a kilometre in length. Blocks of sedimentary and volcanic rocks of Upper Triassic to Middle Cretaceous age are either equivalent to those occurring within the Mamonia terrane or have no known source in Cyprus. In addition, there are sheets of serpentinite that are very similar to serpentinites found in the ophiolite. It is likely that all of these exotic blocks and sheets were emplaced into a clay-rich matrix (represented by the undisturbed Moni Formation) in a deepwater environment during a short-lived episode of sliding down submarine slopes during the late Upper Cretaceous.

Strictly, the Kathikas Formation postdates the docking of the Troodos microplate and Mamonia margin, as it lies across the suture zone between them and is not tectonically deformed. This formation is also the first to have developed after the early palaeo-rotation of the Troodos microplate, which was associated with the collision with the Mamonia margin. The Kathikas Formation consists of clasts of Mamonia rocks of variable size and rarer blocks of Troodos material that lie chaotically in a red and purple matrix of clay, silt and sand. This poorly sorted debrite passes upwards into deepwater calcareous rocks of the Lefkara Formation or is directly overlain by younger units of the circum-Troodos sedimentary succession.

The rocks examined at the two stops described below are representative of the Kannaviou, Moni and Kathikas Formations. In addition, the Kannaviou Formation may be examined at stops 4.4, 5.6, 6.3 and 6.4, the Moni Formation at stop 6.11 and the Kathikas Formation at stop 5.6.

Stop 6.1 The Kannaviou and Kathikas Formations between Agios Dimitrianos and Kritou Marottou   This stop consists of several closely spaced localities for examining the lithologies of the Kannaviou and Kathikas Formations and their associated mass movement. It is worth starting off with an overview of the area from the small church on a hill (36 460139 E, 38 63037 N), which lies on the northern side of the E703 between Agios Dimitrianos and Kannaviou. The concrete access road to the church is probably suitable only for cars.

Outside the church there is an excellent 360° panoramic view of the Ezousa Valley. The background of brown hills to the northeast comprises upper crustal rocks of the Troodos ophiolite. Of note is the recently completed Kannaviou dam, a concrete-faced embankment structure that sits on Troodos volcanic rocks. The Kannaviou reservoir is integrated into the Pafos Irrigation Project by way of a water conveyor to Asprokremmos reservoir. In the valley bottom, to the south of Kannaviou village, there are occasional prominent masses of Troodos serpentinite, which elsewhere are associated with Mamonia rocks. The high ground on the valley sides is in many places capped by white calcareous rocks of the Lefkara Formation. In between the ophiolitic and Lefkara rocks are clay-rich lithologies of the Kannaviou (grey) and Kathikas (reddish-purple) Formations that show obvious evidence of mass movement, and these are now investigated in detail.

Return to the E703 and head up hill (west) to the outskirts of Agios Dimitrianos, and then turn right onto the E712 to Lasa and park almost immediately at the intersection with the road off to the right to Kritou Marottou (36 458901 E, 38 63086 N). The parking place is on the southern edge of a large area underlain by highly mobile clay-rich lithologies of the Kannaviou and Kathikas Formations that extend almost all the way to Kritou Marottou, which itself sits on the Lefkara Formation. The area above and below the road to Kritou Marottou is prone to mass movement, which has given rise to the terraced and hummocky morphology of the landscape. The most obvious feature is the large slump scarp to the north, around which the road to Fyti runs precariously. From this road it is possible to look down on the lobes that typify slumped material that has flowed plastically (Fig. 6.3).

The road to Kritou Marottou resembles a distorted patchwork quilt with its many repaired sections. These, along with the tilted telegraph poles in the area, signify that slopes remain unstable and are differentially flowing and creeping down hill. The movement is exacerbated during heavy rain, because not only do the bentonitic clays swell and flow, but these clays and the silts are rapidly eroded, leaving deep gullies and reducing slope-supporting material. The extensive drainage pattern is very obvious and it attests to the extent of surface flow during rain. However, the problem of slope instability is not restricted to wet weather, because during hot and dry periods the bentonitic clays shrink and effectively subside. The landscape along the road to Kritou Marottou is, therefore, highly dynamic and continues to change annually with fluctuations between wet and dry weather.

Now travel along the road to Kritou Marottou to examine representative outcrops of the Kannaviou and Kathikas Formations. A large coach will have to proceed with caution along the narrow road. Stop at the first obvious exposure on the left-hand side of the road (36 458922 E, 38 63496 N), where volcaniclastic sandstone crops out as one of several relatively resistant ribs in more eroded greyish bentonitic clay. These are lithologies of the Kannaviou Formation, which here lies beneath the Kathikas Formation exposed near by and is examined at the next locality. The

Figure 6.3 View eastwards from the road between Agios Dimitrianos and Fyti, showing the area investigated at stop 6.1 on the western slopes of the Ezousa Valley. The two large lobes in the foreground were formed by mudflows in the Kathikas Formation; the hummocky terrain in the distance is characteristic of the unstable clay-rich Kannaviou and Kathikas Formations. The road between Agios Dimitrianos and Kritou Marottou crosses this unstable ground and the village of Kannaviou sits in the valley bottom.

sandstone is a single thick bed of re-sedimented pyroclastic ejecta and it displays some upward grading from coarser to finer material, the latter exhibiting faint planar laminations. In the lower part of the bed there are flattened pumice fragments and pieces of crystals of amphibole, feldspar and quartz, as well as slightly altered glass shards. These components and the geochemistry of the rocks are indicative of explosive calc-alkaline volcanism.

Carry on along the road towards Kritou Marottou and stop in front of a prominent outcrop exposing dark reddish-purple rocks on the left-hand side of the road (36 459366 E, 38 64473 N). This is the back scarp of a major rotational slump within the Kathikas Formation and it exposes highly disorganized material typical of a debrite. This rock consists of a reddish-purple mudstone and siltstone matrix that hosts angular clasts up to 2 m in diameter, which may define a very faint stratification where they are aligned. The clasts are lithologically diverse, but by far the most abundant rocks are Mamonia chert, sandstone and lava, with Troodos rock types present in much lesser quantity. At this locality the Kathikas rocks rest unconformably on grey Kannaviou clays, and together they are highly unstable.

142

Stop 6.2 The Moni mélange and Lefkara Formation along the E109 south of Parekklisia This stop involves visiting two localities along the E109. The first is at the southern limit of Parekklisia and the second is 1.5 km south, at the intersection with the F122 to Pyrgos. If coming from the A1 motorway, exit at junction 21 and head northwards on the E109 to Parekklisia. Park on the large flat area on the western side of the road, a short distance south of the football stadium (36 514704 E, 38 43639 N).

The view to the south reveals a topography in the foreground and middle distance that is controlled by large competent blocks set within a bentonitic clay matrix. This is the Moni mélange, and its clay component readily creeps, flows and slumps down slope. The higher ground in the far distance comprises rocks of the younger Lefkara Formation that dip south; they are investigated at the next locality. The mélange rests on top of the Troodos ophiolite, and immediately west of the parking area there are exposures of volcaniclastic sedimentary rocks and highly brecciated pillow lavas that belong to the ophiolite.

Perhaps the most notable rock type at this locality is the massive yellowish sandstone that forms the hill to the east of the parking area. This is Parekklisia sandstone within the Moni Formation, and a look north will reveal that it overlies ophiolitic volcanic rocks exposed in shallow roadcuts. A closer examination of the sandstone shows that it is quartz rich, friable and very coarse, generally having angular to sub-angular grains. There are occasional marl-rich bands, containing abundant plant remains, and large concretions cemented by iron oxides. This type of sandstone may be of Lower Cretaceous age and it does not appear to occur elsewhere in Cyprus. Its quartz content probably derives from eroded granite, another rock type that does not occur in Cyprus, but which was probably part of the Mamonia margin. At other locations close by, the sandstone was quarried for the manufacture of coloured glass and for use in the construction and ceramics industries.

Now travel 1.5 km south on the E109, noting how the landscape changes with rock type, and park on the open area on the left (east) immediately after the intersection with the F122 to Pyrgos (36 515028 E, 38 42212 N). Be very cautious here, as the E109 is very busy with fast-moving traffic. From the safety of the parking area it is easy to see that the E109 cuts through highly cleaved, folded and faulted chalk with chert bands of the Middle Lefkara Formation, which are here part of the Yerasa fold and thrust belt (see stops 6.6 and 6.11 for more details).

From a position of safety behind the crash barriers, walk several hundred metres northwestwards up the E109 and note that the Lefkara rocks become less deformed, richer in marl and more pink; the two latter features are typical of the Lower Lefkara Formation. From this elevated position there are good views north and around to the east. The white rocks to the east are more calcareous rocks of the Lefkara Formation, but those on the hill to the north are unconformable reef limestones of the Koronia Member of the Pakhna Formation. The Lefkara and Pakhna rocks overlie the Moni mélange, which occupies the lower ground and contains a variety of rock blocks and sheets in a bentonitic clay matrix. Parts of extensive sheets of Troodos serpentinite crop out in this area. Of particular note is that the serpentinite inside the fork between the E109 and F122 is locally overlain unconformably by umbers of the Perapedhi Formation and by volcaniclastic sedimentary rocks typical of the Arakapas fault zone. This association implies that serpentinite was exposed on the sea floor when the umbers and volcaniclastics were deposited, a conclusion also reached elsewhere (see stop 4.5). All of the rock types seen from the vantage point may be examined by returning to the parking area and walking several hundred metres along the F122 towards Pyrgos.

The Lefkara Formation

The intense tectonic activity associated with collision of the Troodos microplate and Mamonia margin was followed by a quiescent period that allowed for the accumulation of the significant thickness of pelagic carbonates of the Upper Maastrichtian to Oligocene Lefkara Formation. These are best developed to the south and east of the Troodos massif, where they form units up to several hundred metres thick. They are poorly preserved along the western margin of the ophiolite and on the Akamas Peninsula, as here they were either eroded after accumulation or never deposited. It would seem that non-deposition is most likely, as both the Troodos and the Akamas appear to have remained topographical highs throughout the deposition of the calcareous sediments.

The deposits that make up the Lefkara Formation are deep marine and mainly biogenic in origin, and are composed of planktonic forams, calcareous nannofossils and subordinate radiolarians. Traditionally, the formation is divided into Lower, Middle and Upper units. The Lower Lefkara is latest Maastrichtian in age and consists of thin-bedded pinkish-grey marls and marly chalks with occasional chert nodules. It rarely reaches more than 50 m in thickness and is commonly found filling depressions in the volcanic palaeosurface of the Troodos ophiolite, especially where the Perapedhi and Kannaviou Formations are absent. The lower unit is followed by the Middle Lefkara, which is Palaeocene to Eocene in age and comprises up to 500 m of white chalks with extensive chert horizons in its lower portion, and an upper part of massive mainly chert-free white

chalks that pass laterally and vertically into more fissile marl-rich horizons. As the marl content increases, the middle unit grades into the Oligocene Upper Lefkara, which is up to 200 m thick and consists of grey marls
and marly chalks. The complete sequence described here cannot necessarily be seen everywhere because the lithological elements within the
Lefkara Formation are not developed and distributed uniformly. Nevertheless, it provides a useful reference sequence from which to work.

The stops described below enable examination of representative sections of all the Lefkara units, starting in the Lower Lefkara and working
upwards to the Upper Lefkara. In addition to these stops, overviews of the
contacts between the Lefkara Formation and older rock units may be
gained at stops 3.12, 3.14, 3.15, 3.16, 4.4, 5.1, 5.6, 6.1, 6.2 and 7.12. Although
not detailed here, there are several excellent road sections through the
Lefkara Formation, which may be appreciated while driving (carefully);
for example, along the B8 near Trimiklini and the A2 motorway between
Larnaka and its junction with the A1 motorway.

*Stop 6.3 The Lefkara Formation along the E616 between Mandria and Agios
Nikolaos* This stop provides a complete overview of the Lefkara Formation by examining four localities along a 3 km stretch of the E616 (F616
in places) heading west towards Agios Nikolaos from Mandria. It is best
approached in this direction in order to work up-section. The road runs
approximately parallel to the undulating contact between the Troodos
ophiolite and the overlying Lefkara Formation, and it affords many good
near and far views of this contact (Fig. 6.4), and the intervening Perapedhi
and Kannaviou Formations in places.

The traverse begins about 5 km west of Mandria, where there is a cutting on the left (south) side of the road, which exposes lavas unconformably overlain by Lower Lefkara marls and pinkish marls that may contain
minor chert nodules (36 480 506 E, 38 58 440 N). The contact is marked by a
thin zone of greyish-green clay that presumably belongs to the Kannaviou
Formation. The outcrop has yielded the oldest date for the basal Lefkara
in Cyprus. With the aid of a hand lens, millimetre-size rounded forams
may be seen in chert layers in blocks that can be found alongside the road
outcrop. Note that these cherts were probably derived from the Middle
Lefkara, which is exposed higher up the slope.

At a distance of 0.7 km farther along the road towards Agios Nikolaos,
there is a large steep outcrop on the left (south), which has several boulders
at its base (36 479 967 E, 38 58 596 N). Here, lower Middle Lefkara chalks and
bedded and nodular cherts are exposed. Sedimentary structures, such as
cross bedding, parallel lamination and sole structures, are present, as are
trace fossils, including *Thalassinoides*. The chert is secondary in origin and

Figure 6.4 The view westwards from the E616 east of Agios Nikolaos (stop 6.3), showing the contact between whiter rocks of the Lefkara Formation (right foreground and left) resting on top of greyer rocks of the Troodos ophiolite (middle and right).

its distribution within a sequence of chalk beds probably reflects variations in the original content of siliceous microfossils.

Continue 1.1 km along the road. Here, after a bend to the left and before a bend to the right, the road runs straight and has cuttings on both sides. At the end of the straight, just before the bend to the right, park on the right immediately after outcrops on the right and opposite large beds of white chalk with pink-to-red weathering surfaces (36 479181 E, 38 58738 N). These are upper Middle Lefkara massive chalks that pass laterally and vertically into cleaved marly chalk. The interconnecting cleavage is probably diagenetic in origin, formed as a consequence of pressure-solution mechanisms that concentrated insoluble clay minerals into thin films along what are now the cleavage planes.

A further 1.2 km along the road, just before the roadsign indicating straight on and Lemesos to the left, pull over on the left and park (36 478014 E, 38 58386 N). Walk back a short distance to observe the uppermost Upper Lefkara Formation marls in partially vegetated and somewhat poor roadcut exposures. These deposits weather to soft crumbly material when exposed, but in fresh samples from boreholes they are thin to medium bedded and pale to dark grey, and they contain very thin black organic-rich layers. From this point it is convenient and logical to examine rocks of the lower Pakhna Formation at stop 6.8 along the F615 towards Arsos.

146

Stop 6.4 Troodos volcaniclastics and Lower Lefkara marls along the F111 west of Pano Lefkara   At exit 13 on the A1 motorway, take the E105 into Pano Lefkara and then the F111 heading west towards Vavatsinia for about 2 km. Stop at the first contact between white Lefkara rocks and green rocks of the Troodos ophiolite (36 526221 E, 38 58716 N). It is not advisable to park here; safer parking places may be found farther up the road.

The Lower Lefkara marls lie on ophiolitic basement, which is here seen as spectacular submarine volcaniclastic debris flows that are worth examining first. These flows contain well rounded to angular clasts that are dominated by basalt (some vesicular). Within the unit there are in places whole pillows that exhibit chilled margins and radial cooling joints. Farther west along the road, two separate flows can be observed; the contact between them is hummocky and undulating, and the upper unit contains clasts that are much coarser than those in the lower one. There is no obvious stratification in the flows, although there is some faint size grading of the clasts in places.

The debris-flow unit is overlain, in turn, by a 1 m-thick coarse sandy volcaniclastic sediment, a reddish-weathering clay, a green-weathering bentonitic clay (possibly of the Kannaviou Formation) and, eventually, pinkish to grey marl with nodular and bedded chert that represents a typical portion of the Lower Lefkara Formation. The chert varies in colour and is granular to vitreous, and some chert beds preserve cross bedding and planar laminations. Farther along the road to the east there are flattened chert nodules. The basement topography upon which the Lefkara Formation was deposited was uneven, as can be seen by looking to the northeast towards a house and small farm. The contact between the ophiolite and the Lefkara rocks dips into the valley and appears again underneath this farm.

Obviously, the village of Lefkara gives its name to the geological formation upon which it sits, but it is more famous for its lace. It has been claimed that Leonardo da Vinci bought lace for the cathedral of Milan when he visited the village in 1481. In the past the women stayed at home making lace while the men travelled the world to sell it. Lefkara remains the centre of Cyprus lace-making to this day, and also silver is worked for jewellery. It is worth wandering around the picturesque village, as there are lace-making establishments, coffee shops and attractive stone buildings, many of which are rather imposing, attesting to the wealth generated by lace production. From here it makes sense to examine the Middle Lefkara Formation at stop 6.5.

Stop 6.5 The Middle Lefkara chalks along the E105 southeast of Pano Lefkara
The stop examines exposures in roadcuts on both sides of a very straight

section of the E105 that lies approximately equidistant between junction 13 of the A1 motorway and the village of Pano Lefkara. There are no real landmarks to aim for, except a house above the cutting on the southwestern side of the road, close to which it is possible to park (36 530730 E, 38 55235 N). It is logical to combine this stop with stop 6.4.

The roadcuts expose Middle Lefkara chalks and calciturbidites that exhibit beautiful sedimentary structures, particularly on the northeastern side of the road, which include cross laminations (highlighted in black in the more competent beds), load structures and, in places, bioturbation. On this northeastern side the beds dip approximately 20° to the southeast and the less competent beds appear to be marly chalk, but they are not. They are in fact less silicified chalks that are weathering in a lenticular manner, possibly caused by the disintegration of marl-lined ripples that developed on the sea floor (Fig. 6.5). The whole sequence on the northeastern side of the road is cut by a series of normal faults dipping moderately to the northwest, adjacent to which there are some drag folds.

On the southwestern side of the road there is a contact between the interbedded siliceous and less siliceous chalks and a more massive chalk with dark-brown surfaces. The contact has been progressively moved in the up-road direction by a series of shallow northwest-dipping normal faults. This is in contrast to the more steeply dipping faults on the other side of the road, where there is no massive brown-weathering chalk, suggesting that some sort of structural discontinuity runs along the road.

Figure 6.5   The Middle Lefkara Formation, in which bands of silicified chalk stand proud of less-silicified chalk that is preferentially weathering through the disintegration of marl-lined ripples (stop 6.5). The section is about 5 m high.

Within the brown-weathering chalk there is some evidence of grading and there are some intraclasts visible on white surfaces. There are also load structures at the base of the dark-brown weathered coarser beds, as well as evidence of slumping, syn-sedimentary folding and many minor faults offsetting individual beds by several centimetres. These coarser beds exhibit lamination, but no grading.

Stop 6.6 The Middle and Upper Lefkara Formation, Yerasa fold and thrust belt, and a Pleistocene marine terrace at Governor's Beach Governor's Beach lies on the coastal side of the B1 southwest of Mari and is accessed from junction 16/17 (coming from the west) or junction 15 (coming from the east) of the A1 motorway. Follow the clear signs to Governor's Beach and, as the road deteriorates, turn left at the sign for the Kalymnos restaurant and park in the restaurant car-park a short distance beyond the entrance to a campground. The beach is reached by steps behind the restaurant. In order to examine the whole shore section described here, it is best to start on the obvious flat promontory of Middle Lefkara chalk (36 525248 E, 38 41819 N). From here, walk a couple of hundred metres to the southwest, to where there are chert bands within Middle Lefkara chalk, before walking back to the promontory and farther northeast to just beyond a small valley where orange Upper Lefkara rocks are exposed. All of the Lefkara rocks show some degree of deformation that may be attributed to the Yerasa fold and thrust belt (see also stops 6.2 and 6.11).

At the southwestern end of the transect it is clear that there are small thrust faults within the succession and these are well defined by displaced chert bands. Back at the promontory the cherts have highly irregular shapes and orientations, some suggesting boudinage, and there is clear evidence that the fracture cleavage within the chalk and chert has been folded. Thin highly bioturbated organic-rich grey bands within the chalk are offset by small faults. Much of the Lefkara chalk is bioturbated, but it is very difficult to see this in the pure white rock. In the cliff immediately behind the promontory, the Lefkara rocks have been cut into and filled by a Pleistocene channel deposit that is overlain by a Pleistocene marine terrace.

The fracture cleavage is better developed in the cliffs farther to the northeast. On the northeastern side of the valley, the chalks and cherts are highly deformed, with the cherts defining many fold patterns. There is a superb angular unconformity between the contorted and moderately to steeply dipping Lefkara rocks and an overlying horizontal Pleistocene marine terrace (Fig. 6.6a).

The cliffs that meet the water here take on an orange colour and the rocks contain a higher proportion of marl. At first glance the fracture

Figure 6.6 Examples of sedimentary rocks and features at Governor's Beach (stop 6.6). **(a)** Dipping beds of Upper Lefkara chalk and marly chalk overlain unconformably by horizontal beds of sandstone and conglomerate belonging to a Pleistocene marine terrace. **(b)** Trace fossils in Upper Lefkara marly chalk. The coin is 28 mm across. (Both images kindly supplied by Charlie Underwood)

cleavage does not appear to continue upwards into the orange units, but closer inspection will reveal that it does, and its preservation is probably controlled by the competency of the rock types. These Upper Lefkara chalks and marly chalks contain an abundance of well preserved trace fossils, such as *Zoophycos* and *Thalassinoides*, which are especially clear on wave-washed surfaces, and which again demonstrates the high degree of bioturbation preserved in these sedimentary rocks (Fig. 6.6b).

The marine terrace seen in the cliff along much of the beach section is clearly unconformable with the underlying Lefkara rocks. The unconformity is flat on the large scale, but can be seen to be irregular in detail (Fig. 6.6a), especially where it overlies hard beds of chert. The terrace deposits comprise variably cemented sandstones and conglomerates containing large boulders of chalk, chert and ophiolite-derived material. Fossils are abundant, and all belong to species that still exist, indicating a relatively recent age for the terrace (probably Pleistocene, about 190 000 years old). Fossils are dominated by bivalves, such as *Glycymeris*, with gastropods and the scaphopod (tusk shell) *Dentalium* also present. Some of the carbonate boulders contain borings of the bivalve *Lithophaga*. As this deposit represents the 8–11 m (above sea level) terrace, it demonstrates that this part of Cyprus has risen relatively little in 190 000 years compared with other areas such as western and southwestern parts of the island.

On return to the clifftop, the view to the northeast reveals the large industrial complex at Vasiliko, which contains at least one cement factory. Its location is close to sources of raw materials for cement making: greyish marls provide clay and can be seen close to the works, whereas the surroundings hills provide an abundance of calcareous rocks (for the production of lime) and gypsum.

The Pakhna Formation

The Miocene saw the development of several basins that coincided with significant uplift of the Troodos massif (see Fig. 6.1). The basins allowed a range of different environments to develop, which resulted in accumulations of highly variable sediments that make up the Pakhna Formation. Within the basin centres there was relatively little change in conditions from the deposition of the Lefkara sediments, so that hemipelagic sedimentation dominated. Closer to the basin margins, however, turbidites and other remobilized sediments were much more abundant. Overall, the Pakhna Formation records a progressively shallowing marine environment during the Miocene, which ended with the crystallization of evaporites of the Kalavasos Formation.

The contact between the white Lefkara Formation and the younger cream rocks of the Pakhna Formation is usually very sharp and is marked by a sudden influx of turbiditic calcarenites and the deposition of hemipelagic chalks intercalated with silty marls. This contact may be

diachronous and it has been shown to be either conformable or unconformable, depending upon location. The thickness of the Pakhna Formation is comparable to that of the Lefkara Formation, but it was deposited
over about 17 million years, as opposed to 50 million years for the latter.
This rapid sedimentation reflects substantial uplift and erosion of the
Troodos massif during this time. Indeed, the Pakhna Formation shows
increasing terrigenous (ophiolitic) input up sequence and includes
organic-rich layers derived from plant material eroded off an emergent
Troodos massif. Because of the non-uniform uplift and erosion of the massif, significant variation exists in the volume, content and nature of deposited sediment, and this makes it very difficult to subdivide the Pakhna
Formation into a formal vertical lithostratigraphy. Nonetheless, throughout the formation, superb large- and small-scale sedimentary features are
preserved (Fig. 6.7).

Reef facies are present at various levels within the Pakhna Formation,
but are most common at the bottom and top of the formation, where they
are referred to respectively as the Terra and Koronia Members. The Terra
Member is evidence that sea depth was very shallow in places, even in the
Lower Miocene. The presence of ophiolitic debris in sediments of the same
age adjacent to the Limassol Forest suggests that some land had emerged
at this time. At the top of the formation, and of similar age to the Koronia
Member, is a very useful marker in the form of an interval containing the
bottom-dwelling marine foram *Discospirina*, which lived in shallow water
no deeper than 50 m. This unit is Upper Miocene (Upper Tortonian) in age
and its deposition appears to have been continuous into the evaporites of
the Kalavasos Formation.

The stops presented below are arranged in approximate sequence from
bottom to top of the Pakhna Formation. As mentioned previously, lithologies and structures within the formation are highly diverse, making it
impossible to define a type section. The fullest appreciation of the Pakhna
Formation will, therefore, be achieved by visiting as many of the stops as
possible. Most of the stops are located to the south of the Troodos massif,
where the Pakhna rocks are most extensively preserved. In addition to the
stops detailed below, the Pakhna Formation may also be examined at
stops 5.1, 5.2, 6.12, 7.9, 7.10 and 7.15, and on the way to stop 6.11.

Two road sections are worthy of mention, but they have not been
described in detail. The A6 Lemesos–Pafos motorway cuts through spectacular faults, folds, slump structures, intraclasts, channel fills, unconformable contacts, erosive surfaces, reef structures and gypsum of the
sedimentary succession – but do not stop on the motorway to observe
them. Similarly, along the F614 between Kouklia and Dora there are some
magnificent road exposures of the Pakhna Formation, which is folded and

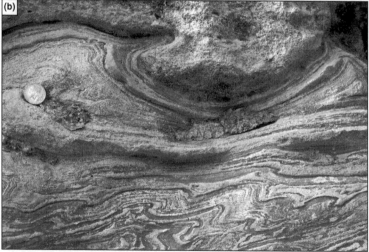

Figure 6.7 Examples of sedimentary features in the Pakhna Formation. **(a)** Bedded calcarenite and marl (F614 northeast of Pano Archimandrita: 36 471854 E, 38 46080 N). The lower half of the section is dominated by a bed of marl that contains large clasts of highly contorted and isoclinally folded bioclastic calcarenite that slumped into the marl. The height of the section is 1 m. **(b)** Syn-sedimentary deformation features (stop 7.9). The upper part of the image shows a load cast and an associated oblique flame structure where coarser-grain calcarenite overlies finer-grain marl. In contrast, the lower half of the image shows contorted or convolute lamination in marl that contains thin bands of calcarenite. The coin is 28 mm across. **(c)** Extensively bioturbated bioclastic horizon, in which burrows are well preserved (stop 7.15).

faulted, and contains superb slumps, load structures and intraclasts (Fig. 6.7a). On the same road, the area around Pano Archimandrita exposes sections through the Kannaviou, Lefkara and Pakhna Formations.

Stop 6.7 The Terra Member reef limestones and Quaternary sandstones in the Cape Gkreko area   This stop is well worth visiting, despite being at the southeastern tip of the island and a long way from any other. The Cape Gkreko area has been designated one of the natural beauty spots of Cyprus and it is easy to spend up to a day here. The area offers swimming, diving, cycling, riding, climbing and excellent hiking. From Agia Napa, head east on the E306 towards Cape Gkreko before turning onto the E307 towards the cape. About 3 km before the cape, on a bend around to the left, there is a sign for a viewpoint and nature trail; turn right (south) at this sign and follow the road (F313) along to a parking place in front of the viewpoint. Walk south along a rough track towards an obviously quarried face that lies below the viewpoint (36 597410 E, 38 69507 N). It is advisable to wear a hard hat in the quarry, as the quarry faces are unstable.

The quarry has been cut into a patch reef of the Lower Miocene Terra Member of the Pakhna Formation and it provides one of the best exposures of an ancient reef in Cyprus. The reef limestones seen here are composed of a framework of coral colonies, especially of poritids and faviids that may individually reach well over 0.5 m across (Fig. 6.8). The corals are associated with laminated structures that were formed by layers of calcareous red algae. With patience, a wide variety of corals and other reef organisms and structures may be observed.

154

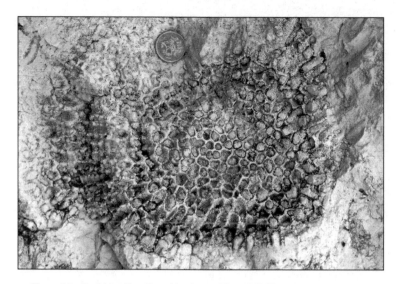

Figure 6.8 Faviid coral in a Terra Member reef (stop 6.7). The coin is 25 mm across.

Very obvious features against the white background are the near-vertical green veins that cut the reef framework. These veins contain angular fragments of white limestone and are infilled with chert, marl and sand, which in places exhibit some internal banding that is usually parallel to the vein margins. The veins are clearly younger than the reef and must have formed after the initial lithification of the limestone. It has been suggested that veining was initiated during crustal extension prior to the Pliocene. As the veins dilated, they became progressively infilled by different sediment, the marls and sands perhaps representing a component of the Nicosia Formation deposited during a Lower Pliocene marine transgression that submerged the reef. Vein propagation and dilation may have been aided by movement on a bentonitic mélange that is known to underlie the Lefkara Formation, upon which the Pakhna Formation rests conformably in this part of Cyprus.

From the quarry, walk up to the viewpoint (36 597249 E, 38 69353 N), from where there are stunning vistas of the surrounding area. The viewpoint is about 100 m above the sea, and the hill upon which it is sited is composed of reef limestones that formed in shallow sea water in the Lower Miocene. This gives some idea of the amount of uplift experienced by this southeastern corner of Cyprus since that time. The viewpoint itself was once the site of a Neolithic settlement and, later, of a sanctuary dedicated to Aphrodite.

The viewpoint affords a good view to the east of the radio masts on

Cape Gkreko itself, which is where to head next to examine Quaternary beach sandstones. Return to the junction with the E307, turn right (east) and travel a short distance before turning right (southeast) again onto the F314 to Cape Gkreko. Travel to the end of this road and park in front of the fenced area of radio masts (36 598734 E, 38 69532N).

To the north and south of the parking area there are bioclastic marine sandstones exhibiting magnificent cross bedding; the largest foresets may have been calcareous sand dunes. These Quaternary sands have been cemented into a type of beachrock and uplifted, and shells within them are the same as those found on the modern beach.

Stop 6.8 The lower Pakhna Formation along the F615 between Agios Nikolaos and Arsos The stop lies along an extensive roadcut on the eastern side of a straight section of the F615 between Agios Nikolaos to the north and Arsos to the south (36 477990 E, 38 57112 N), and is about 1.5 km south of where the F615 joins the F616 on the outskirts of Agios Nikolaos. It is logical and convenient to examine this stop after stop 6.3.

The cutting exposes a strikingly well bedded and gently folded sequence of the lower Pakhna Formation (Fig. 6.9). The sedimentary rocks comprise cleaved and marly chalks interbedded with yellowish calcarenites, which are typically turbiditic. With time and patience, it is possible to observe a multitude of sedimentary features, including load structures, planar laminations, concretions and slumps, and many trace fossils (such as *Zoophycos*). A flaggy cleavage, with planes separated by thin films of

Figure 6.9 A gently folded, well bedded sequence of marly chalks, calcarenites and minor marls of the lower Pakhna Formation (stop 6.8).

clay, is well developed in many horizons and is typical of the lower Pakhna Formation. This cleavage is probably diagenetic in origin, arising from the expulsion of clay minerals during pressure-solution recrystallization of the rock.

Farther south from this location the Pakhna rocks change. Towards Arsos the proportion of marl increases significantly, and around Malia, Pachna (stop 6.9) and areas still farther south, the rocks become more sandy. These changes reflect sediment deposition in a progressively shallowing marine environment.

Before leaving, view the valley below to the south, west and north, which is filled with rotational-slip blocks of lower Pakhna rocks similar to those examined in the roadcut. Rotational failure occurred during the Quaternary, as continued uplift of the Troodos massif triggered movement within Upper Lefkara marl that underlies the Pakhna Formation here and is exposed on the F616 to the north (stop 6.3). The slipped blocks are zones of groundwater recharge, and springs at their bases allowed villages such as Agios Nikolaos and others in the valley to become established.

Stop 6.9 Changing Pakhna facies north of Pachna village   This stop follows on logically from stop 6.8 and involves a traverse of about 1 km along the main road immediately north of Pachna village, from which the Pakhna Formation takes its name. The road section is accessed from the north by taking the turning towards Pachna from the E601. If coming from the south, leave the A6 motorway at the junction for Avdimou and drive north through Pachna. It is best to start at the northern end of the section and walk southwards, up hill towards the village. Start about 400 m southwest of the intersection with the E601, where there is an outcrop on the southeastern side of the road, opposite a track and a bench (36 481789 E, 38 50205 N). There is parking at both ends of the traverse.

At the base of the outcrop, relatively massive calcarenite is overlain by marls, which dominate the rest of the exposure. The thinly bedded to laminated grey and brown marls (and possibly marly chalks) contain several beds of calcarenite, which exhibit load structures. Farther up hill, the marl-rich horizons are overlain by orange-weathering calcarenite. The alternation between marl- and calcarenite-rich horizons continues along the section, with overall more marl than calcarenite.

About 600 m from the start of the traverse there are cuts on both sides of the road that expose interbedded white chalk, orange calcarenite and greenish cleaved marl (36 481537 E, 38 49668 N). The marl contains a high proportion of sand, but the most distinctive feature is the abundance of trace fossils. Within the marl there are heavily burrowed intraclasts of calcarenite and chalk, which must have been burrowed and lithified before being ripped up.

About 400 m farther on, there is a track on the western side of the road, along which there are coloured (white, buff, tan and orange) beds of sandstone with and without appreciable amounts of marl. Again, these rocks are highly burrowed, but they also contain abundant shell fragments. The buff and orange beds contain plant remains and possibly ophiolitic detritus, as their colour suggests the presence of iron-bearing minerals.

Overall, on moving southwards and up hill along the road, the exposures record the deposition of different sediments, probably associated with a progressively shallowing water depth, and extensive biological activity. The plant remains suggest that parts of the Troodos massif must have been above sea level at the time of sediment deposition.

Stop 6.10 The Discospirina *bed and channel deposits of the uppermost Pakhna Formation at Choirokoitia*   Two localities in the uppermost Pakhna Formation are included in this stop, which respectively examine the *Discospirina* bed and channel-fill deposits. This stop may be conveniently combined with stop 7.5. From the A1 motorway, take exit 14 and follow the F112 as it skirts around the southwestern side of Choirokoitia village. Park on the left by the old church, about half-way through the village. Continue along the road on foot, watching for traffic, as far as the sharp right-hand bend and obvious cutting on the inside of this bend. This bend is very dangerous, so highly visible clothing should be worn. A safer alternative is to observe the outcrop from a viewpoint on the opposite side of the road (36 530445 E, 38 50527 N).

The roadcut exposes thin- to medium-bedded chalks, marls, limestones, sandstones and siltstones of the uppermost Pakhna Formation. At the base is a layer of bivalve-bearing hemipelagic chalk, which contains *Discospirina*, a flat spiral-ornamented benthonic foram up to 6 mm in diameter. To find it, use a hand lens to examine loose fragments of the bed – there is no need to hammer. This marker bed was probably deposited in a maximum water depth of 50 m and is traceable over about 27 km, from Kalavasos to the west, to near Kiti in the east. The *Discospirina* bed is overlain by laminated orange-to-buff sandy siltstones and shales that are rich in gypsum and exhibit occasional cross bedding. The gypsum is both concordant and discordant with respect to bedding and at least some of it may be secondary. The laminated beds are overlain by a 40 cm-thick coarse bioclastic layer containing bivalves, forams and overlapping tabular intraclasts that exhibit sharp lower and upper margins. Above this layer there are pale-grey marls, medium-grain calcarenites, gypsum and, finally, reef limestones in the northernmost part of the roadcut. The whole sequence was probably deposited in shallow marine waters and is indicative of the transition between marine sedimentation and evaporitic conditions associated with the uplift of the Troodos massif.

Continue approximately 300 m farther along the F112, going down hill, before turning onto the dirt road on the right (northeast). Almost immediately, to the right, is a turnoff up to a parking area. From here continue on foot to where the road ends at a T-junction. Turn right and walk about 100 m up towards the village of Choirokoitia, to a track off to the left. Walk down slope along the track and take a left at the first split in the track, then almost immediately, at the second split in the track, take the right track and follow it all the way down to the valley bottom of the Maroni River. Cross the river bed and continue a short distance up the track to where there is an excellent view to the southwest of a cliff, on top of which is the village (36 530757 E, 38 50951 N). The total distance from the parking area to the cliff is about 500 m.

To its left and right the cliff is composed of khaki marly sands and silts; in the centre these are cut by a large channel filled with conglomerate (Fig. 6.10). The fill of the channel is composed of several thick units separated by flat bedding surfaces and by thin beds of marl. The continuation of these marl horizons into the rocks hosting the channel implies that what

Figure 6.10 Part of the uppermost Pakhna Formation, in which marly sands and silts (only partially visible in the bottom left-hand corner behind the treetop) are cut by a series of stacked channels each filled with a horizontal unit dominated by conglomerate (stop 6.10). The lower five channels are shown in this figure, and the large boulder of Lefkara chalk (about 2 m in diameter) is part of the fill of the third channel. The thin horizontal marl-rich horizon either side of the top of the boulder appears to continue into the rock hosting the channel.

appears to be one large single channel is probably a stacked series of channels that were eroded and filled while the fine khaki sediment was accumulating.

Suitably protected by a hard hat, walk over to the cliff and examine the marly sands, which contain thin beds of limestone. The marls are dated as Tortonian (Upper Miocene) from the delicate white pteropod fossils they contain, and are clearly marine. They contain black grains of ophiolitic debris and abundant plant remains, showing that, in the source region of the sediment, the ophiolitic rocks of Troodos were already exposed and vegetated. The clasts in the channel fill may be examined in debris at the base of the cliff; they include boulders of Lefkara chalk and ophiolitic rocks, and rounded lumps of *Tarbellastraea*, a coral common in the Upper Miocene. As the channel and the marls are of the same age, the channel fill must also be Tortonian in age. With reference to the previous (*Discospirina*) locality, limestone and gypsum must overlie the marly sands and silts exposed here in the cliff.

For those interested in investigating more rocks of the uppermost Pakhna Formation, there are some thick beds of calcareous sandstone exposed in quarry faces in a valley running northwest away from the lovely old stone village of Tochni (36 529449 E, 38 49087 N), which lies south-southwest of the present stop.

Miocene uplift of the Limassol Forest

Units within the Pakhna Formation clearly demonstrate that parts of the Troodos massif had emerged above sea level and were being eroded in the Middle and Upper Miocene (e.g. stop 6.9). The Middle Miocene was also a time of significant uplift and associated deformation of rocks in the Limassol Forest, and serpentinites and gabbros in this area are now exposed at high elevations, partly as a consequence. The western and central portions of the Limassol Forest were uplifted to a greater extent than the eastern portion. Associated thrusting was directed in a radial pattern away from the centre of the main serpentinite blocks in the western Limassol Forest, and was focused along pre-existing lineaments.

At the time of uplift and thrusting in the Middle Miocene, the Limassol Forest area lay within an overall compressive regime because of continued northward subduction beneath Cyprus; serpentinite already occupied an elevated position and was exposed on the sea floor as a consequence of transform-fault deformation in a spreading environment. The position of the Limassol Forest with respect to the Troodos massif at this time meant that rocks in the former were pushed against the more competent sheeted dykes of the massif to the north, and this resulted in uplift and southward escape (thrusting) of rocks in the Limassol Forest. A component of the uplift and thrusting may also be attributed to hydration and expansion associated with the pervasive serpentinization of peridotite.

One of the most impressive features arising from the structural re-organization was the Yerasa fold and thrust belt, which is up to 5 km wide and runs northwest–southeast along the southwestern edge of the Limassol Forest (see Fig. 6.1). Here, deformation was directed to the southwest, with folding in the adjacent Lefkara and Pakhna sedimentary rocks a result of their compression during movement of the adjacent serpentinite. The Lefkara Formation in particular exhibits upright or southwest-facing folds and intense axial planar cleavage close to the thrusts. Between the ophiolitic and Lefkara rocks there may be a zone of sheared and folded Moni clays or mélange, conspicuous because of their deep erosion and slumping, and these are well exposed at stops 6.2 and 6.11 in the Yerasa fold and thrust belt. In addition to these stops, another part of this belt may be examined at stop 6.6.

Stop 6.11 The Yerasa fold and thrust belt along the F128 south of Akrounta
This stop is a transect that begins towards the northern end of the Germasogeia reservoir and runs for about 1 km north along the F128 to a bridge over the Monastiri River immediately south of the village of Akrounta. Coming from the south, exit the A1 motorway at junction 24 and head north on the F128. There is a very good exposure of Pakhna rocks by the war memorial on the western side of the F128, a short distance north of the A1. Travel through Germasogeia, past the dam and a superb outcrop of breccias and conglomerates opposite a narrow embayment in the reservoir, and park on the right (east) side of the road almost immediately after it cuts through another deposit of conglomerate. The parking area also marks the start of a roadcut on the left (west) side of the road (36 507804 E, 38 45625 N). Equally, this parking spot may be reached by travelling about 1 km south from the outskirts of Akrounta, where there is ample additional parking if needed.

Start by examining the roadcut opposite the parking place. If rocks of the Lefkara Formation have been examined previously, it will be clear that the chalks, marls and chalky marls exposed here are noticeably different. They exhibit an excellent fracture cleavage, which is at a high angle to the bedding (Fig. 6.11) and generally strikes northwest–southeast, the same as the trend of the main structural features along the Yerasa fold and thrust belt. For a fuller appreciation of the regional structure and distribution of deformation, walk about 1 km along the road to the outskirts of Akrounta, taking care because of fast traffic. Along the way the proportions of marl

Figure 6.11 Steeply inclined fracture cleavage cutting flat-bedded Lefkara chalks, marly chalks and marls (stop 6.11). The lower beds are up to 30–40 cm thick.

and chalk vary, as does the intensity of the fracture cleavage. In some places cleavage is so extreme that the rocks are highly brecciated, with fragments of more competent chalk set in a marl matrix. It is worth determining if the composition and hardness of the beds have any influence on the amount of cleavage they exhibit, and whether the intensity of cleavage within a given rock type changes along the road section. It is also important to keep a note of the strike and dip of the beds and cleavage along the route.

The last exposure before the river and Akrounta is composed of massive chalk of the upper Middle Lefkara Formation, in which the bedding is very steep to vertical as a consequence of folding. Overall, these chalks are dipping to the northeast, which means that somewhere along the road section at least one fold axis has been crossed. At this point, the view across the small embayment reveals more very steeply dipping highly cleaved beds of chalks and marls.

From the bridge (36 507658 E, 38 46417 N) it is clear to see that there is a sharp break in slope trending approximately northwest–southeast. This marks the contact between the Lefkara rocks and the red, brown, orange and grey bentonite-rich Moni mélange upon which parts of Akrounta sit. The mélange occupies a valley bounded to the immediate north by hills made up of sheared and brecciated volcanic rocks, dolerite and microgabbro of the Limassol Forest. From the bridge it is possible to appreciate that the huge mass of serpentinite lying a short distance away in the hills to the northeast has been thrust southwestwards during its uplift. In so doing, it caused southwest-directed thrusting of other ophiolitic slivers and the Moni mélange, and compressed the Lefkara Formation, leading to the very pronounced folding and fracturing near its contact with the mélange. Rocks of the Pakhna Formation that lie to the southwest were also affected by the compressional forces.

The Kalavasos Formation

During the latest Miocene (Messinian), about 5–6 million years ago, the Mediterranean Sea became isolated from the Atlantic Ocean and was in the grips of a salinity crisis. As evaporation took place, waters turned hypersaline and extensive deposits of evaporite minerals formed. The cause of the crisis is not fully understood, but it may have resulted from either a fall in relative sea level, perhaps related to regional tectonic uplift or a major glacial event, or formation of a barrier between Spain and Morocco because of movement of the African plate against the Eurasian plate.

In Cyprus, the Messinian salinity crisis is recorded in the Kalavasos Formation, which comprises gypsum alternating with marly chalk, chalky marl and minor silt. This alternation implies that there was cyclicity in the rise and fall of sea level during the Messinian. The sporadic occurrences of gypsum generally reflect the distribution of the basins in which they formed, many of which had been developing throughout the Miocene (see Fig. 6.1). The Mesaoria, Polemi (part of the larger Polis basin) and Pissouri basins developed through extensional faulting that caused graben to form, whereas the Maroni basin developed between pre-existing compressional lineaments created by localized thrusting.

The gypsum deposits are generally attractive and photogenic (Fig. 6.12), but their many textures make them difficult to understand, with each texture reflecting a specific depositional setting or post-depositional process. In simple terms, these evaporites are likely to have formed in small and somewhat isolated basins, in which the maximum brine depth was probably less than 100 m. Very coarse-grain gypsum (selenite; Fig. 6.12a) crystallized *in situ* in shallow water (no deeper than 10 m) at the basin edge or within very shallow lagoons. Also developed in this marginal environment, but as a relatively minor facies, was a sugary granular anhedral type of gypsum. It may have grown at the brine/sediment interface or below the subsurface during early diagenesis, or it may be clastic in origin and derived from pre-existing selenite. In contrast to these

Figure 6.12 Examples of gypsum in the Kalavasos Formation. **(a)** Massive crystalline swallowtail gypsum in growth position on top of marl (stop 6.13). The maximum height of the gypsum block is 2.5 m. **(b)** Finely laminated and alabaster gypsum (gypsum quarries south of Tochni: 36 529442 E, 38 47015 N). The coin is 25 mm across. **(c, opposite)** Lenses of secondary selenite in primary laminated gypsum (stop 6.13). **(d, opposite)** Radiating crystals of selenite growing off laminated and alabaster gypsum, and developing upwards into a breccia of selenite fragments in a marl and silt matrix (stop 6.12). **(e, overleaf)** Slump folds in laminated gypsum that overlies relatively undeformed sugary-texture banded gypsum (gypsum quarries south of Tochni).

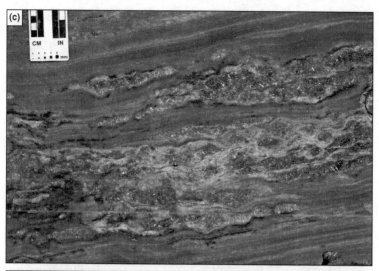

marginal facies, fine-grain gypsum precipitated in the surface water across the entire basin and sank to accumulate on the basin floor. These accumulations may have been influenced by bottom currents that led to the development of finely laminated (laminae usually 1–3 mm thick) gypsum known as marmara, or they may be more massive (alabaster). Marmara and alabaster are usually intercalated (Fig. 6.12b), and together they represent the most extensively preserved gypsum types on the island.

Primary gypsum facies were at times modified by later processes that

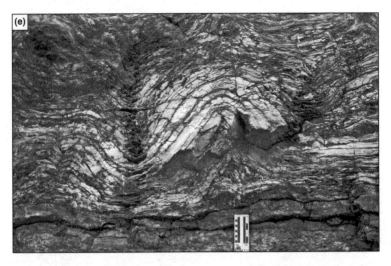

caused recrystallization or disruption (Fig. 6.12c,d,e). Marginal deposits of selenite were often destabilized by fault movement and earthquakes, and consequently slumped and flowed into deeper parts of the basin, sometimes producing chaotic and brecciated masses that also disrupted other pre-existing deposits. Clasts of algal stromatolites may also occur in this disturbed gypsum and these are either pre-Messinian (being derived from the Koronia Member) or a contemporary basin-margin facies.

The stop detailed below is located in the Maroni basin. Stop 6.13 examines gypsum in the Polis basin, which contains some of the most spectacular deposits in Cyprus. There are many gypsum types and relationships at these two stops.

Stop 6.12 Gypsum and reef limestone at Maroni From the A1 motorway, take exit 14 or 15 and follow the signs to Psematismenos and Maroni. On leaving Psematismenos take the first left turning signposted to Maroni. Note that the village centre of Psematismenos is too narrow for a large coach. After entering Maroni, take the first paved left turn onto a road with new houses. Drive straight up the hill and park where the tarmac ends. Continue straight up on the dirt road, follow it around a 90° bend to the right and walk up to the large reef knoll on the left, just before the crest of the hill and a rusty fence (36 532637 E, 38 46885 N). From here it is best to explore the slopes down from the road, where varieties of gypsum deposited in the Maroni basin are associated with reef knolls. Do not hammer any exposures and, please, collect loose samples only if absolutely necessary.

The relatively flat gypsum pavement exposes a wide variety of gypsum types and relationships, not all of which are easy to describe or under-stand. An examination of the entire surface and a shallow quarry on the eastern side will reveal laminated marmara and bedded alabaster (Fig. 6.12b), gypsum breccia and conglomerate, minor massive sugary gypsum, and selenite rosettes, domes, veins and quasi-dykes. In general, the grey and cream marmara and alabaster, in places containing lenses of second-ary selenite formed by *in situ* recrystallization of the finer gypsum (Fig. 6.12c), are overlain by gypsum breccia, and also occur within it as large patches. The marmara and alabaster are steeply inclined where domes of selenite rosettes and breccia poke through the pavement surface. These rosettes are probably secondary in origin, produced by recrystallization during compaction of the marmara and alabaster. Gypsum breccia is also found away from the domes, but it may have been derived from them as domes grew up, fragmented and were eroded by water currents, perhaps during storms. The rounded appearance of many of the selenite fragments indicates some degree of reworking before or during final deposition, and the occurrence of chalk clasts suggests the addition of material from an external source. In places, the alabaster gypsum is fractured and intruded by the breccia.

Crystals of radiating milky-white selenite are abundant and some may be seen growing off laminated alabaster gypsum in the shallow quarry faces (Fig. 6.12d), but it is not clear whether they are related to dome growth. Others occur in the breccia in veins and dykes that reach widths of 1.5 m and may crosscut one another. In these bodies, it would appear that gypsum nucleated on vein or dyke margins and grew inwards. The abundance of fractures, veins and dykes in, and filled by, gypsum, suggests that perhaps these gypsum deposits were affected by fault and seismic activity during their formation. It is possible that some gypsum is intru-sive, formed by the injection of gypsum-saturated brine-rich pore waters expelled during compaction of pre-existing gypsum, perhaps aided by groundshaking related to seismic activity.

Several reef knolls are present and vary in diameter from less than a metre to several tens of metres or more (Fig. 6.13). The age of these knolls with respect to the gypsum is not entirely clear. Some appear older (prob-ably part of the Koronia Member) than the gypsum, but others are con-temporaneous with it. On the western edge of the area, there are several small irregular carbonate bodies integrally related to the gypsum, but at the first knoll below the road it appears that the gypsum laps onto the knoll and is younger, although one part of the knoll appears to be younger by overlying the gypsum. This knoll has a rubbly base, but it rapidly becomes massive and finely laminated throughout. The laminae are indicative of algal mounds that grew in warm shallow clear water, but the absence of

Figure 6.13 Rubbly looking reef knoll with gypsum pavement in the foreground (stop 6.12) (image kindly supplied by Charlie Underwood).

other reef organisms, such as coral, suggests that the water was highly saline and not suitable for normal reef growth. A close examination of the knoll reveals abundant white crystals of celestite lining voids in the limestone. This sulphate mineral is secondary in origin and most probably precipitated from pore fluids that had previously leached strontium from microbial carbonate enriched in this element.

The presence of reefs knolls closely associated with gypsum suggests that much of the gypsum was deposited either in shallow lagoons or on the margin of a deeper basin, both of which were protected from the sea by reefs. However, these environments may be disputed, as the presence of breccia, marmara and alabaster could indicate the deeper parts of a basin, with the breccia representing a debris-flow deposit. In order to be more specific, an investigation of the surrounding area would be required.

Those interested in examining gypsum deposited in another, possibly deeper, part of the Maroni basin should investigate the quarries adjacent to the F116 south of Tochni (36529442 E, 3847015 N), which lie about 3.5 km due west of the present stop. Permission is required to gain access to the working quarries.

Pliocene and Quaternary sedimentation

The Messinian ended with the development of palaeosols and the occurrence of brackish and fresh waters in some areas. By contrast, the Messinian–Pliocene transition was marked by a very rapid input of sea water into the Mediterranean basin as it was reconnected to the Atlantic Ocean through the modern opening at the Strait of Gibraltar. In Cyprus, therefore, the Pliocene was a period of renewed marine transgression, with plankton-rich fine marl sedimentation in fault-controlled basins and coarser sand, sandy marl and conglomerate deposition in shallower waters around areas of positive relief. These sediments and sedimentary rocks belong mainly to the Nicosia Formation, which is well developed, for example, in the Polis basin to the west of the Troodos massif and the Mesaoria basin to its north (see Fig. 6.1).

The Quaternary was a time of renewed and drastic, although episodic, uplift of the Troodos massif. This was accompanied by rapid erosion and denudation to form widespread immature coarse chaotic sediments, called fanglomerates, which blanket most of the older rocks (Fig. 6.14a). In south and west Cyprus, uplift is documented by a series of marine terraces that become progressively older upwards (Fig. 6.14b). The uppermost terrace can be seen along the highest ridge of the Akamas Peninsula and is the last in a series of five exposed levels, each related to a pulse of sea-level change superimposed on the overall tilting and uplift of this part of the island. Even today, beachrock is forming in the coastal zone around Cyprus and some of this will be uplifted above sea level in the future (Fig. 6.14c). The record of sedimentation and emergence during the Quaternary is complicated, because this period saw large variations in sea level associated with global glacial and interglacial periods.

The erosion of Pliocene and Quaternary sediments and rocks in places gives rise to a distinctive landscape, which is particularly well developed on the Mesaoria Plain south and southwest of Lefkosia. The abundant marls and brown silty sands and sandstones of the Nicosia Formation erode to produce a spectacular badland topography, which is highly unstable, prone to movement and very difficult to build on. However, these materials do stand proud in mesas capped by gently dipping conglomerates, usually Quaternary in age and of the fanglomerate unit, that protect the underlying marl from erosion (Fig. 6.14a).

In addition to the stops detailed below, Pliocene and Quaternary rocks and sediments can be examined at stops 5.9, 6.6, 6.7, 7.7, 7.9, 7.15, 7.16 and 7.17, and on the way to stop 6.11. Although not documented in detail here, Holocene beachrock and Plio-Pleistocene sandstone may be examined on the beach immediately to the southeast of Pafos airport (36 454271 E, 38 40686 N), and these deposits may be compared with similar ones at stop 7.9.

Stop 6.13 Messinian evaporites and Pliocene sedimentary rocks along the F622 between Eledio and Amargeti This stop in southwest Cyprus relates to localities along 1.4 km of the F622, which examine the evolution of the Polis

Figure 6.14 Examples of Pliocene and Quaternary sedimentary rocks and landforms. **(a)** Mesa developed where Pleistocene fanglomerate caps extensively gullied marls and silts of the Pliocene Nicosia Formation (view east towards stop 6.14 on the road crest). **(b)** Faulted marine-terrace deposits at about 100 m above sea level, which consist of interbedded paler calcarenites and darker fluvial–marine boulder conglomerates (sand and gravel quarry in Agia Varvara village: 36 455338 E, 3845728 N). The largest boulders are about 50 cm in diameter. **(c)** Gently dipping Holocene beachrock developed in the zone of wave splash (stop 7.9).

basin, starting in the Messinian Kalavasos Formation outside Eledio and ending in the Pliocene Nicosia Formation at Amargeti to the northeast. Coming out of Eledio, the road first bends to the left, then to the right and then to the left again. Between these last two bends there is a white concrete road off to the left, on which it is advisable to park (36 460776 E, 38 52315 N). Do not hammer any exposure and, please, collect loose fragments only if really necessary.

On the northern side of the F622 are slope and quarry faces exposing near-horizontal gypsum, which is dominated by marmara and alabaster (see Fig. 6.12b). Inspection of the exposure to the left (northwest) will reveal several metres of brownish-grey to cream finely laminated marmara and bedded alabaster gypsum. Crystals within the laminated gypsum are elongate and lie parallel to the lamination plane. Within this primary gypsum are lenses of secondary (diagenetic) grey to white coarser-grain selenite, in which the crystals show no preferred orientation (see Fig. 6.12c). The contacts between the laminated gypsum and selenitic lenses are usually marked by sugary gypsum, which, like the selenite, is also a product of recrystallization of the laminated gypsum.

Now climb up onto the flatter top of the exposure, which is underlain by marl. In front, part way across a field, is a spectacular outcrop consisting of, from bottom to top, a mixture of marl and patches of selenite, about 0.5 m of marl, and a 2.5 m-high block of massive crystalline selenite (see Fig. 6.12a). The latter appears to be in growth position, having a flattish contact with the marl below and subvertical crystals of selenite. These crystals are metres long and they exhibit superb twins, giving the selenite a swallowtail structure.

The whole section examined here probably represents a shallowing basin, from deeper laminated facies to marl and selenite deposited in a shallow lagoon or along a basin margin. The laminations within the basal gypsum are 1–3 mm thick and may represent varves that formed in the Polemi basin, a sub-basin of the larger Polis basin (see Fig. 6.1). From a high vantage point in the area it is possible to look to the northwest and see a section through the Polis basin, on both sides of which the beds dip in towards the basin axis.

More time may be spent at this locality to examine the gypsum further. Gypsum breccia occurs near by and there are shallow excavations resulting from the removal of gypsum for use as floor tiles. Many of the walls in neighbouring villages have been constructed from the laminated gypsum because it breaks readily into rectangular blocks.

Now travel 500 m farther towards Amargeti, to where a section of the old road can be seen on the left (west) side of the road (36 461190 E, 38 52410 N). Park here and walk down the old road about 50 m to an outcrop consisting of clusters of giant elongate swallowtail gypsum crystals, each crystal several metres long and about 10 cm wide. These crystals contain brown bands every 30 mm or so. If the bands represent breaks between annual growth, then the crystals grew at least ten times faster than the deposition rate of the laminated gypsum (laminae 1–3 mm wide) at the previous locality. The swallowtail selenite may have formed rapidly in local hollows and depressions on the basin margin or in a shallow lagoon. Destruction of this selenite by wave action, probably during storms, provided the material to feed the coarse gypsum breccias that may be seen at the previous locality.

Continue 600 m northeastwards along the F622, to a high white roadcut on the left (western) side of the road (36 461510 E, 38 52679 N). Between here and the last locality the roadcuts are dominated by gypsum and marl, but here the roadcut exposes younger (Nicosia Formation) marine chalk containing abundant and remarkably preserved forams. In places, the chalk is covered by hillwash, which must be excavated.

After a further 300 m along the F622 towards Amargeti, there is a turning to this village and a bus shelter at the junction (36 461785 E, 38 53000 N). The road section northeast of this junction exposes material that lies above the previous marine unit. Almost all of this material is terrestrial and subaerial in origin, consisting of fluvial gravels interbedded with white calcareous hillwash material, although one of the white units towards the southern end of the roadcut contains forams similar to those at the previous locality. In the middle of the eastern roadcut there is a fault with a throw of several metres, and north of this are smaller faults with the fault blocks all thrown down to the southwest. The faults are clearly synsedimentary, because the fluvial gravels thicken on the downthrown blocks, and they must be young, although there is no exposed fault scarp in the vineyard above the roadcut.

This last roadcut lies at the centre of the southern end of the Polis basin and is about 400 m above sea level. The sediments preserved in the cut represent late-stage basin fill.

Stop 6.14 Pleistocene fanglomerate overlying the Nicosia Formation along the E907 between Arediou and Episkopeio This stop lies about 20 km southwest of Lefkosia city centre and examines outcrop where the E907 cuts through the crest of the obvious ridge between the villages of Arediou to the west and Episkopeio to the east (Fig. 6.14a). It is possible to pull off and park a car on the northern side of the road on the western side of the ridge, where the most informative section is exposed (36 520171 E, 38 78268 N).

The lower part of the cutting exposes a series of grey and brown marine marls with siltstone bands that belong to the Nicosia Formation. These give rise to the extensive badland topography of the surrounding countryside, but they may stand proud in mesas capped by fanglomerates, such as those along the ridge through which the road cuts (Fig. 6.14a).

The Nicosia Formation is overlain unconformably by the fanglomerate unit, which consists of a sequence of lenses of poorly sorted conglomerate within brownish marl. The lenses, representing cross sections through channels, generally have flat upper surfaces and concave, sometimes irregular, lower surfaces. The channel fill is dominated by sub-rounded pebbles, but angular clasts do exist, and there are large boulders up to 0.5 m in diameter. Many pebbles are cracked, indicating that deposition occurred in a high-energy environment, and are mainly ophiolitic (lava, microgabbro, dolerite and diabase). These would have been washed off

173

the rising bulk of the Troodos massif immediately to the south.

Within the section there is a distinctive orange to red palaeosol on top of the fanglomerates, but it has been suggested that it is not *in situ* and is in fact upside down, because its sharp bright-orange base is out of place. If *in situ*, its base would be expected to be gradational with the underlying fanglomerate, and its top would have the most intense orange colour. The palaeosol was probably ripped up from top to bottom during a flood event and then deposited here top first and bottom last.

The entire section records a change from a shallow-marine to a fluvial environment as the Mesaoria basin was uplifted, along with the Troodos massif to the south, and emerged above sea level. Although it is not described in detail here, a similar record exists in an excellent cliff section at the site of Chrysospiliotissa, the shrine of Our Lady of the Golden Cave (36 525761 E, 38 83186 N), near the village of Deftara to the northeast.

Stop 6.15 Pleistocene environments recorded along the Kouris River southeast of Episkopi   Along the B6 west of Episkopi is the turning to the archaeological site of Kourion (stop 7.15). A few hundred metres east of this turning, take the E602, signposted to Akrotiri, from the B6. Follow the E602 for 2 km, first heading south before bending around to travel east, to just before the bridge over the Kouris River. Here, turn left (north) onto a track and park on the right after about 50 m (36 492120 E, 38 35170 N). Walk from the parking place around the eastern side of the mound ahead to find the north–south section of interest, about 300 m long and up to 20 m high, which looks over the cobbles of the Kouris River bed. Unfortunately, this locality is at times used as a dump, which can make it rather unpleasant.

The section permits examination of a complex sedimentary record of marine, fluvial and soil-forming environments that existed during the Pleistocene. It is too complex to describe in full here, but can be unravelled by identifying some key features. First, locate the marine units. The cream calcareous outcrop close to the southern end is certainly marine, as it contains oyster shells and forams. Similar marine beds can be found at the base of the section in the centre of the outcrop and are also present at the very northern end, where they occur above a fluvial unit. However, there are other units of fine-grain cream sediment that cannot be shown to be marine, as they do not contain identifiable marine fossils. These units may be calcrete or caliche.

Next, identify the fluvial conglomerate units. These look very similar to the present river bed and they contain rounded cobbles of a variety of ophiolitic lithologies. About half-way along the outcrop is a cross section through a spectacular small fan that progrades out into, and is partly interbedded with, a unit of cream calcareous sediment. The conglomerates in this fan contain forams and may represent a submarine shoreface deposit. Finally, note the reddened soil horizons, underlain by nodular calcrete, which show the effects of prolonged exposure to air.

The section is interpreted to be the product of a complex interplay between uplift of the island and fluctuating sea level during the Pleistocene. A generally continuous uplift has raised the products of marine, fluvial and soil environments, which formed at times of varying sea level as glaciations lowered the sea to expose land, and interglacial events caused a rise in sea level and submergence of land.

From here it is convenient to visit Kolossi castle, which lies on the southern outskirts of Kolossi village. To reach it, return to the road and head east for about 2.5 km before turning left (north) and travelling a further 1 km. The present keep-style structure dates from about AD 1454 and stands on the site of a much grander structure that was built in about AD 1210, when the Lusignans granted the site to the Knights Hospitaller (the Knights of the Order of Saint John of Jerusalem). From AD 1310 the Hospitallers maintained Kolossi as a commandery and this is where the rich dessert wine Commandaria takes its name. Wine and sugar-cane production in the area brought the knights considerable wealth.

Stop 6.16 Pleistocene fluvial boulder conglomerate along the F612 northeast of Kouklia From the B6 between Timi (west) and Pyrgos (east), take the turning marked by brown signs to the Sanctuary of Aphrodite and the Palaia Pafos museum (both of which are worth visiting and have been designated world heritage sites by UNESCO). Head into Kouklia and then travel about 700 m northeast towards Archimandrita on the F612. The outcrops of interest are on the right (southeast) side of the road opposite the sign for leaving Kouklia, where there is plenty of parking (36 461479 E, 38 41115 N).

The roadcut exposes spectacular river-channel deposits overlying marly chalk of the Pakhna Formation (Fig. 6.15). The top of this chalk is bleached and hardened to a depth of about 20 cm and this probably represents a formerly exposed and weathered surface on which calcrete developed prior to deposition of the overlying river sands and gravels. The channel deposits are stacked, and later ones cut down into earlier ones. The deposits grade upwards from the base of coarse ophiolitic (gabbro, microgabbro and dolerite) pebbles and boulders to sand at the top. There is plenty of evidence of cross bedding and cross stratification, and the sandy horizons exhibit bedding and lamination. Sandy material must have been reworked, as there are clasts of this in the channel fill.

Not far down the road to the southwest, the channel fill contains more carbonate clasts. At least some of these appear to be derived from the

Figure 6.15 Pleistocene river channel deposits overlying vegetated marly chalk of the Pakhna Formation (stop 6.16).

underlying bedrock, which must have been scoured and eroded as the channel cut down into it, and such features are suggestive of erosion during flood events. The sandy horizons are also well developed and show well defined bedding with cross lamination.

These channel deposits are Pleistocene in age and are associated with the 100 m marine terrace. The village of Kouklia also sits on such a terrace, but a slightly lower and therefore younger one. This flat younger terrace, with its commanding position overlooking the sea, was exploited by the ancients: it is where the once great city of Pafos (Palaia Pafos) was originally located. For a time this city was the greatest of Aphrodite's shrines, but it declined during the Roman period because of repeated earthquake damage over several centuries. Two earthquakes in the fourth century AD caused the capital to revert from Palaia Pafos back to Salamis (Constantia). The demise of Palaia Pafos came with the Arab raids in the seventh century AD.

Stop 6.17 Pleistocene facies change and modern salt deposits at the Larnaka salt lakes To the south of Larnaka, and to the north and south of its airport, are several salt lakes, the largest of which is examined here. To reach the appropriate part of this lake, follow directions to Larnaka airport and then take the A4 towards Larnaka and Lefkosia from the roundabout on the outskirts of the airport. Travel northeast several hundred metres along the southeastern edge of the lake, and park in the first layby (36 557308 E,

176

38 60851 N), which is opposite a concrete water tank. Walk about 10 m down to the lakeshore (bearing in mind this is the shoreline only after a winter of very heavy rain), keeping to the left of a pile of rubble. From this point on the shore (36 557285 E, 38 60851 N) examine the facies change in the sedimentary rocks along the shoreline for about 250 m to the northeast, finishing at a small bluff just below the old road (36 557485 E, 38 61034 N).

Immediately to the southwest of the access point are some small exposures, on the shore, of a buff muddy sandstone. When the lake level is unusually high, this exposure can be partly covered. The sandstone is calcareous and it is likely that much of the sand is actually shell fragments. Scattered small pebbles of ophiolitic rocks are present, along with some fossils, mostly small gastropods. It is the trace fossil *Ophiomorpha* that is most abundant, its characteristic pelleted burrow lining being very obvious. The low diversity of fauna and the predominance of *Ophiomorpha* may suggest deposition within a somewhat restricted environment, possibly an estuary or a slightly hypersaline lagoon.

About 20–30 m to the northeast there are exposures of a conglomerate of varied ophiolitic pebbles in a carbonate sand matrix. This conglomerate contains a very high-diversity assemblage of shelly fossils (Fig. 6.16a). Many of these fossils are highly fragmented, whereas others are simply separated at the joints and otherwise well preserved. All of the species present appear to exist today, although not all are now present in the sea around Cyprus. Probably the most conspicuous are the colonies of the coral *Cladocora*, which are in life position, overturned or fragmented (Fig. 6.16b). Others include many gastropods and bivalves, and rarer scaphopods (tusk shells), serpulid worm tubes and echinoids. Many of the ophiolitic pebbles, and some of the larger shells, are coated in a pale calcareous crust produced by the red alga *Lithophyllum*. This indicates shallow-water conditions, and the complete coating of many pebbles suggests that energy levels were high enough to have turned them during the growth of the algae. The presence of rare examples of the large gastropod *Strombus bubonius* (conch) suggests warm-water conditions. This species entered the Mediterranean Sea during a very warm interglacial period at about 175 000 years ago. It is no longer present in these waters, now being restricted to the subtropics off West Africa, although shells of a smaller species, *Strombus persicus*, have been found on some beaches in Cyprus. This is an Indian Ocean species that was first recorded in the Mediterranean in 1983, having entered through the Suez Canal.

On moving farther along the shoreline to the northeast, sporadic exposures of the conglomeratic sandstone occur, in which some large coral colonies may be seen. On reaching the small bluff (which may be accessed from the main road if the lake level is too high), the conglomeratic unit

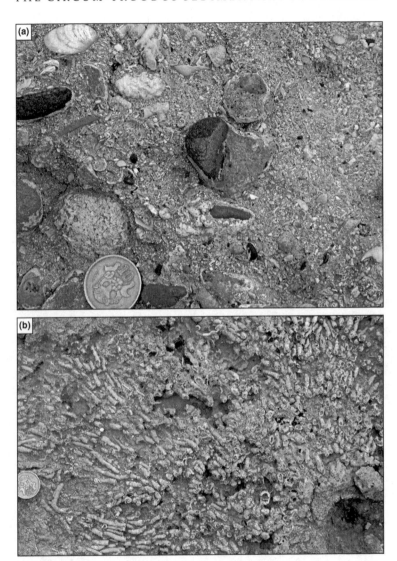

Figure 6.16 Pleistocene sedimentary rocks and fossils at Larnaka salt lakes (stop 6.17): **(a)** pebbles coated by a pale calcareous crust and set in a coarse bioclastic sand matrix; **(b)** colony of the coral *Cladocora*. The coin is 25 mm across in both photographs. (Both images kindly supplied by Charlie Underwood)

fines upwards to a fine carbonate sand. Fossils are uncommon, but thin-shell bivalves, scaphopods and rare echinoids are present, some in life position in their burrows. The top 1 m of the bluff is heavily calcretized and difficult to interpret.

Taken together, the shoreline exposures appear to show a general deepening trend to open marine conditions, from what were probably restricted marine or estuarine conditions. The marine facies show a change from shallow-water conglomerates at the base and grade up into finer, more offshore, sands. The rocks are Pleistocene in age and they formed the foundation of a barrier between the sea and the lake, against which younger Pleistocene and Holocene gravels and sands were deposited. The chemical similarity of the lake and sea waters suggests that the salt lake results from infiltration of sea water through the barrier of loosely consolidated and permeable gravels, sands and sandstones. The lower level of the lake relative to sea level creates a hydraulic gradient through the barrier that drives infiltration into the lake all year round. During summer months the rate of evaporation exceeds the rate of infiltration, and salts form as the lake dries up. During winter and early spring the saline lake supports brine shrimp, which every year attract thousands of pink flamingos and other migratory birds.

The Larnaka salt lakes are classic coastal salinas. Their layers of evaporite salts even caught the interest of Pliny the Elder and, later, the Venetians, who generated significant state revenue by exploiting them. The lake bed is about 2–3 m below sea level and is covered by black mud with a high organic and salt content. The salts crystallizing on top of this mud layer are mainly halite and much smaller amounts of sylvite, magnesium salts and gypsum. The salt was mined by simply lifting it off the underlying mud and allowing it to drain and dry out. It was then piled into large pyramids that stood outside for at least a year to permit rain to wash out the more soluble and bitter tasting potassium and magnesium salts from the halite. The purified halite was used mainly as table salt, but also for pickling olives. Mining recently ceased because of increased contamination of the lake.

To the west-southwest across the lake is Hala Sultan Tekkesi, a mosque erected in 1816 by the Turks during the Ottoman period; it ranks as one of the holiest places of Islam. It marks the spot where the aunt of the prophet Mohammed was buried following her death during an Arab raid in AD 649, when she was attacked by Byzantine forces and broke her neck when she fell from her donkey.

Stop 6.18 Quaternary marine and lake environments recorded in the Akrotiri salt lake   This stop is a half-day transect into the eastern margin of the Akrotiri salt lake, which is great fun as well as very instructive. To the northwest of Lemesos, leave the A1 motorway at junction 29 and head for the port. Follow the signs to the port through two sets of traffic lights. Turn right at the third set of lights and follow brown signs to Lady's Mile Beach. Turn left at the next lights and then drive to a set of lights just by the wall of the port. Turn right here, then left at a stop sign, continuing around the port wall. The road becomes a dirt track after a short distance and turns south to run parallel to the shore of Akrotiri Bay. Drive 4 km from the stop sign mentioned above, passing the Captain's Cabin café on the left and a cluster of four radio masts on the right, until the Lady's Mile café. Park just beyond this (36 500457 E, 38 30059 N). Be aware that, after storms and strong winds, sand drifts may make the track impassable.

The transect involves getting rather dirty, so wear shorts and old footwear that can get muddy, salty and otherwise abused. If shoes or boots are worn, they will have to be removed in order to go barefoot for the last part of the transect. Take a knife or paint scraper, or equivalent, a transparent plastic bag, and matches or a lighter.

The transect illustrates modern environments (marine shore, salt flat and salt lake) and environments of the recent past (dating from when the salt lake was part of a shallow marine environment). The area of the lake is the emerging part of one of the sedimentary basins so characteristic of Cyprus. Perhaps only a few hundred years ago, the area was covered by a shallow sea. At this time the southern end of the Akrotiri Peninsula would have been an island. Gradually the sea floor has emerged and in the future the salt lake will disappear. As it does so, the fill of the sedimentary basin will start to erode and be reworked longitudinally into the marine portions of the basin lying off shore. Such a situation has already occurred in the Mesaoria basin, which emerged above sea level some 1–2 million years ago, and its axis now lies several hundred metres above sea level.

The nature of the transect varies throughout the year, because the salt lake has an annual cycle. During the summer the rate of evaporation is greater than the rate of recharge through the sediments separating the lake from the sea. The level of the lake therefore drops and eventually it dries out to expose an expanse of white salt flats. In winter, sea water accumulates in the lake and all of the salt dissolves. In winter and spring the lake provides an important stop-over for migrating birds, especially flocks of flamingos.

Walk down to the seashore, a few tens of metres from the parking place, and from here walk back westwards for about 1 km to observe some of the following features, which will vary according to the time of the year:

- Modern shore sediments, sand dunes and a sandy shore. The dunes rise several metres above the level of the salt lake and support a flora that is not necessarily salt resistant.
- Sand flats. These make up a large part of the exposures on the transect. Dig with a knife into the sediment. In some places it is brown sand and in others (especially in ruts) the sediment contains a black layer that

smells of rotten eggs, which is characteristic of hydrogen sulphide gas. This gas signifies the presence of sulphate-reducing bacteria. The flora here includes glassworts and other salt-tolerant genera.

- Carbonate-cemented hardgrounds bored by bivalves (*Petricola*). These are relics from the time when the environment was marine. Cementation has produced beachrock, as is seen at many modern beaches in Cyprus, and bivalves have bored into this rock.

- Spreads of cockle shells, which are another marine relic.

- Cyanobacterial mats. Where these occur, the surface layer of the sand flats has been cemented by cyanobacteria. These mats are slimy and jelly-like when wet, but when dry they crack into flakes that roll up into curls. In most places the mats are underlain by a black sulphide-rich layer produced by sulphate-reducing bacteria. The cyanobacteria live by photosynthesis, producing organic matter that they use to reduce sulphate in sea water. Excess hydrogen sulphide is consumed in the precipitation of iron sulphide, which is what makes the sediment black.

- Salt flats (in late summer and autumn) are covered with skeletal hopper crystals of salt, which have developed faces and empty interiors that reflect very rapid crystal growth.

- Salt lake (in spring and early summer), which is a few tens of centimetres deep and has a very soft fluffy carbonate mud on its bottom. When this mud is stirred, bubbles of gas stream up from it. To collect the gas, fill a plastic bag with water and let the gas bubble into it until there is a good pocket of gas in the bag. If successful in lighting the gas, it should burn with a blue flame, indicating methane. This hydrocarbon gas is produced by microbial activity in the sediment of the lake. The organic matter in the lake supports brine shrimp, red copepods and even small fish.

Chapter 7

Resources, wastes and hazards

Introduction

Cyprus lay at the centre of the ancient world and consequently it has played a fundamental role in the history of humankind and Western civilization. With its abundance of natural resources and strategic position at the crossroads of Europe, Asia and Africa, Cyprus has seen the waxing and waning of many civilizations and has a history that goes back some 10 000 years. Civilization first began to flourish during the eighth millennium BC, and since then the Mycenaeans, Achaeans, Phoenicians, Assyrians, Egyptians, Persians, Romans, Crusaders, Venetians, Ottomans and British have all occupied the island and have left their marks (Table 7.1).

Geology and geomorphology have both been prominent in the human development of Cyprus, particularly in dictating the location, prosperity and decline of settlements. Readily available natural resources were of paramount importance, with fresh water the most essential. All communities required a permanent source of such water, which was usually provided by a perennial spring or water-transfer network (Fig. 7.1). Water remains the most vital resource on the island, but stores of it are becoming depleted because of over-use and changes in climate.

Cyprus owes much of its status and prosperity to its geological resources, as, for its size, the island is blessed with a disproportionate abundance and variety of metallic and industrial minerals, and of construction materials. The ancients generated much wealth from the exploitation of metallic minerals, especially those containing copper, but today emphasis has shifted to industrial minerals and construction materials, particularly to meet demand created by road, dam and housing projects. With such exploitation and development comes the inevitable byproduct of waste. Owing to the longevity of human occupation, the current urban development and the huge annual influx of tourists, there are large volumes of waste distributed across the island in disposal sites, both controlled and uncontrolled. In certain cases this waste is threatening the quality of already-vulnerable water resources.

Table 7.1 Brief history of Cyprus (after Cyprus Tourism Organisation 2001).

Period	Major events and features	Sites described
Neolithic Age 8000–3800 BC	Civilization develops along north and south coasts Circular dwellings Stone and, later, pottery vessels Major earthquake around 3800 BC	Choirokoitia (stop 7.5)
Chalcolithic Age 3800–2300 BC	Settlements mainly in western Cyprus Discovery and exploitation of copper Development of copper tools	Lempa (stop 7.6) Skouriotissa (stop 7.3)
Bronze Age 2300–1050 BC	Copper trade brings great wealth Mycenaeans and Achaeans from Greece Greek language, architecture and traditions imported First city kingdoms established	Palaia Pafos (stop 6.16)
Iron Age 1050–750 BC	Violent earthquake around 1050 BC Ten city kingdoms Cult of the goddess Aphrodite flourishes Phoenicians settle at Kition Period of prosperity in eighth century BC	
Archaic period 750–475 BC	Assyrian, Egyptian and Persian control Wealth from east–west trade routes	Agia Varvara-Almyras (stop 7.1)
Classical period 475–325 BC	Cyprus unified into a leading political and cultural centre in the Greek world Rule of Alexander the Great (333–325 BC)	Amathous (stop 7.9)
Hellenistic period 325–58 BC	Rule by the Ptolemies of Egypt (294–58, 47–31 BC) City kingdoms eliminated New capital at Nea Pafos	
Roman period 58 BC to AD 330	Major earthquakes in first centuries BC and AD Apostles Paul and Barnabas visit Nea Pafos (AD 45) Conversion to Christianity	Kourion (stop 7.15) Nea Pafos (stop 7.16)
Byzantine period 330–1191	Cataclysmic earthquakes in fourth century New cities arise and capital moves to Salamis (Constantia) Spread of Christianity as the official religion Arab raids	Panagia tou Kykkou (stop 3.8) Panagia tou Araka (stop 2.10)
Frankish (Lusignan) and Venetian periods 1191–1571	Crusades and Knights Templar (1191–1192) Ammochostos one of the richest cities in Near East Lefkosia made Lusignan capital Venetians (1489–1571) fortify island	Kolossi castle (stop 6.15) Saranta Kolones (stop 7.16)
Ottoman (Turkish) period 1571–1878	Liberation of Orthodoxy Spread of Islam Taxation	Hala Sultan Tekkesi (stop 6.17)
British period 1878–1960	Annexation from Ottoman Empire in 1914 Taxation Increase in Greek population Massive exploitation of metallic sulphides Move for union of Cyprus with Greece (enosis and EOKA) Constitutional change	
Republic of Cyprus 1960–present	Independence Turkish invasion (1974) United Nations mandate Full membership of European Union (2004)	

Figure 7.1 The Kamares aqueduct, built in 1747 and used until 1930, adjacent to the B5 west of Larnaka (36 554647 E, 38 63483 N).

Geological processes, topography, climate and weather have at times had a negative impact on humans and a profound influence on the location and longevity of settlements in Cyprus. As a seismically active and rising island that contains an abundance of potentially unstable slopes and clay-rich lithologies, and one that experiences periodic extremes in weather, Cyprus is prone to earthquakes, relative changes in sea level, mass movements, floods and droughts. These hazards have occurred throughout the human occupation of the island, and will continue in the future. They are well documented in the geological record, the landscape and archaeological finds.

The story of Cyprus copper

No book on the geology of Cyprus would be complete without considering the exploitation of copper, a metal mined on the island since at least the third millennium BC and one that brought great wealth and status to the ancient inhabitants. Indeed, such has been the importance of copper that the orange-yellow colour of Cyprus on the island's flag is there to signify the important link with this metal since antiquity. Even today, experts still argue over whether the name of the island was derived from the word "copper" (the Greek for copper ore is *kupros* and the Latin for copper is *cuprum*) or vice versa.

The main metallic-ore deposits in Cyprus are the massive copper-bearing sulphides in the Troodos ophiolite. These deposits occur almost exclusively in the volcanic sequence and they group into five main mining districts: Limni, Skouriotissa, Tamassos, Mathiatis–Sia and Kalavasos. Almost all of the orebodies ranged in size from 50 000 to 20 million tonnes prior to mining and they contained 0.3–4.5 per cent copper and generally less than 0.2 per cent zinc. It is estimated that about a million tonnes of copper has been extracted over the past 5000 years, mostly in the twentieth century during large-scale mining operations.

The presence of copper ore at or very close to the surface is usually marked by vari-coloured gossans and oxidized stainings, and no doubt these attracted the ancients to the deposits. At that time, Cyprus was extensively forested and the ancients noticed that native copper was associated with zones of oxidized sulphide in forested areas. This copper was produced naturally by pine resin reducing copper sulphate in acidic solutions produced by oxidation, and this was most probably the first copper to be used on the island sometime during the Chalcolithic (literally meaning copper–stone) Age. The discovery of native copper prompted extensive exploration of gossans by the digging of pits, shafts and galleries. According to radiocarbon dates obtained from charcoal fragments in slag piles, mining and smelting of copper ore were well under way by 2760 BC. This is supported by finds of grooved hammerstones, used for metalworking, which date back to the Chalcolithic and Early Bronze Ages. The Mycenaeans and Phoenicians developed the copper industry, but production peaked during the Hellenistic and Roman periods and petered out in Medieval times.

There was a major revival in mining in 1921 at Skouriotissa and it spread rapidly to other areas, culminating in the digging of large open pits to remove huge volumes of ore. Mining declined in the middle to late twentieth century, as ore grade declined and stiff competition grew from enormous copper deposits in the Americas and elsewhere. Mining in the twentieth century was rarely in areas that had not previously been exploited underground by the ancients, which was testimony to the ability of the ancient prospectors and miners.

The most compelling evidence for the extent of ancient copper production is the 40 or so large slag heaps dotted about the Troodos foot-hills. Although many have been damaged or destroyed in the production of road metal and other aggregate, the remnants are now protected by the Department of Antiquities. These heaps have been estimated to have contained some 4 million tonnes of slag, which were waste from the production of 200 000 tonnes of copper during the initial 3000–3500 years of copper exploitation on the island. The oldest slags tend to be earthy and brown, red or orange in colour (Fig. 7.2a); for example, the Mycenaeans used a manganese flux (possibly umber) that turned their slags a brown to reddish-

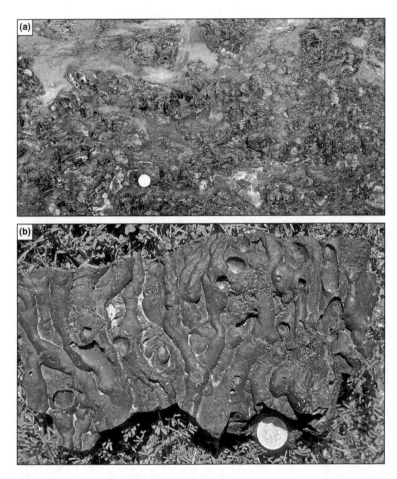

Figure 7.2 Ancient slags: **(a)** earthy slag mixed in with sandy sediment, probably dating from Phoenician time (stop 7.3); **(b)** vitreous black Roman or younger slag (stop 7.4). The coin is 28 mm across.

brown colour. In contrast, vitreous black slags tend to be Roman or younger (Fig. 7.2b). The slag heaps are made up of layers and bun-shape masses, and these may be found close to areas of mineralization, which implies that smelting was undertaken in small furnaces close to the extractable ore. Analysis of the metal content of slag piles has shown that their copper content usually decreases upwards, suggesting that copper recovery improved over time and with experience.

Perhaps the most profound impact of ancient copper production is not immediately obvious today, but it becomes apparent when one considers that Cyprus was once covered by dense forest. In order to smelt copper ore, huge amounts of wood were required for charcoal fuel, and this came from the forests, only a small proportion of which remain on the island. George Constantinou, former director of the Cyprus Geological Survey Department, has estimated that the production of 200 000 tonnes of copper in 3000–3500 years probably required the deforestation of an area at least 16 times that of Cyprus, which was equivalent to deforesting the entire island completely once every 200 years. This was possible because the forests were able to regenerate rapidly enough to keep pace with the fuel demand. Many other copper-producing areas outside Cyprus could not compete, as they were unable to maintain the necessary fuel supply.

The ancients mainly mined for copper, but, during the twentieth century, emphasis shifted to sulphur because of the low metal content of the sulphidic ores. Although copper continued to be produced, vast amounts of pyrite were extracted as a byproduct of mining copper ores, and this pyrite was used extensively in the manufacture of sulphuric acid and sulphide pulp. Cyprus pyrite was highly desirable because it had a high iron content, low concentrations of impurities, especially arsenic, and good roasting qualities. Such was the demand for this pyrite that, at the peak of mining between 1950 and 1970, Cyprus was the third largest producer of pyrite, after Japan and Spain, and second only to Spain in terms of exports. Prior to this, prompted by the mining depression of the early 1930s, there was a decade of gold and silver exploitation, which ended in the mid-1940s. Although gold occurred in many zones of the orebodies, it was usually the gossans that were the most productive. Gold had to be recovered by cyanide leaching of finely milled ore, because grains of this metal were too small to be recovered effectively by mechanical means.

The mining of sulphide deposits during the twentieth century has left its mark on the landscape in the form of abandoned buildings, piles of tailings from ore processing and, above all, 20 or more abandoned open pits containing acidic lakes rich in dissolved minerals and surrounded by large piles of sulphide-bearing spoil (Fig. 7.3a). The pit waters are unlikely to pose a major environmental threat, because there does not appear to be significant connectivity between pit lakes and the water table, thus limiting the possibility of groundwater contamination. The same cannot be said of the piles of tailings and spoil, components of which may be redistributed over considerable distances by wind and water to contaminate soil, sediment and water. The main problem is created when oxygenated waters migrate through the highly porous and permeable piles and oxidize pyrite to sulphuric acid, which may then react further with a variety of minerals. The resulting mineral-rich acidic solution is known as acid mine drainage (Fig. 7.3b). In Cyprus it is fortunate that, around the sulphide mines and mineral processing sites, the alkalinity of the volcanic and

Figure 7.3 Acid mine drainage at the Mathiatis mine (stop 3.15): **(a)** acidic water and sulphide-bearing sediments at the bottom of the open pit; **(b)** sulphate minerals crystallizing on the sides of a drainage channel containing evaporating acidic mineral-rich water.

carbonate sedimentary rocks is capable of neutralizing the acid waters, but the resulting precipitates may be enriched in undesirable elements and these may be remobilized during rain.

Taken together, the stops described below provide a comprehensive overview of the processes and waste products of copper exploitation in

Cyprus. Further investigation of the problems associated with sulphidic wastes can be undertaken at stops 3.15 and 7.12. In addition to the tailings pile described at stop 7.4, there is another excellent example at the Limni mine (east of Polis), but this is not described here. This pile lies 5 km west-northwest of Kynousa, adjacent to the F737 (36 451001 E, 38 79423 N). The geology of the copper-bearing sulphide deposits is covered in detail in Chapter 3, specifically at stops 3.15, 3.17 and 3.18, with additional coverage at stop 7.12.

Stop 7.1 Ancient mining and smelting site south of Agia Varvara The ancient mining and smelting site south of Agia Varvara village is known as Agia Varvara-Almyras and, although it is not spectacular, it is somewhat evocative and worth visiting because of its great archaeological importance. The site, which also lies about 2 km east of the Mathiatis mine (stop 3.15), is reached from the E103 between the villages of Agia Varvara and Mathiatis, and this road may be joined from junction 8 of the A1 motorway. A short distance southwest of Agia Varvara, take the narrow paved road heading southeast from the E103 between the brickworks and a prominent hill capped by a concrete water tank. Within about 500 m the road degrades to dirt adjacent to a factory and is no longer suitable for cars or coaches. From here the site is just over 1 km away. Follow the dirt road sharply around to the right at the next intersection (now heading south) and continue around the base of the prominent brown hill rising towards the east, which is mineralized in places. After passing a wire-fenced construction site, take the left fork where the road splits, and continue around the base of the hill for a short distance until there is a track heading up hill on the left (north). In a short distance this track ends at the mining and fenced-off smelting site (36 533913 E, 38 70552 N). This site is poorly protected, so please do not walk over the fenced-off area and do not take anything away.

The Agia Varvara-Almyras site is so far the only place discovered in Cyprus where an entire copper-production process is preserved (mine, roasting and smelting furnaces, slag, copper ore, refined copper metal and tools). Excavation of the site began in 1988 and pottery finds, metalworking units and radiocarbon dates from charcoal indicate activity at the site to between the seventh and second centuries BC.

Copper ore was mined from the hard silicified gossan 30 m to the northwest of the smelting area. Broken chunks of ore were reduced to fist-size fragments that were then carried to the smelting area to be reduced further by crushing and grinding, using stone implements. Copper-bearing minerals were picked out by hand and transferred to a roasting furnace that was used to reduce the sulphur content of the ore, as too much sulphur

189

hinders the reduction of the ore to metal copper and renders the copper unsuitable for smithing. An elongate roasting furnace has been excavated and found to be orientated towards the southwest, the direction of the prevailing wind. After roasting, the ore was transferred to a smelting furnace, four of which have been unearthed. All were built on bedrock and are quite small, less than 1 m in both height and diameter, and were open at the top to permit recharge with ore and charcoal. The air required for smelting was introduced via handworked bellows and blowpipes. When the liquid copper and slag separated, the latter was collected in a tapping pit in the ground and the copper was allowed to solidify before being removed, which involved destroying the front of the furnace. The raw copper was further refined in workshops off site.

The smelting site sits on volcanic rocks of the Troodos ophiolite, but immediately to the west these rocks are faulted against lavas and dykes of the Basal Group. A steep fault zone, trending northwest away from the smelting area, separates pillow lavas to the southwest from Basal Group to the northeast. The Basal Group along the fault is brecciated, bleached, stained, silicified and mineralized, and has the appearance of a typical gossan. Copper mineralization is closely associated with the fault zone, but the only significant copper deposit in the area occurs in the gossan 30 m to the northwest of the smelting area, where it is marked by an adit into the hillside. This adit is presumed to be ancient, but it may have been re-used during the first half of the twentieth century. Evidence of prospecting activity can be found up slope at other gossans along the fault zone. In all the gossans, pyrite is the main sulphide mineral, occurring as disseminations and massive veins, and in veinlets with quartz. Chalcopyrite is associated with pyrite, but reaches high concentrations only adjacent to the smelting area, where other copper minerals have also been seen.

Stop 7.2 Ancient copper mines along the E616 west of Mandria   Ancient copper mines are located in sheeted dykes of the Troodos ophiolite about 1 km west of a small church on the western edge of the village of Mandria, which lies about 4 km southwest of Pano Platres. They are holes exposed in a cutting along the E616 Mandria–Agios Nikolaos road, and the easiest to see and examine lie about 50 m east of a café and builder's yard (36 483586 E, 38 59040 N).

The holes are the cross sections of adits, probably Roman in age, that appear to follow very sparse malachite staining in fractures in sheeted dykes. These adits run both parallel and at high angles to the dykes, and are most evocative if you imagine (do not try) squeezing into a hole, crawling along with mining tools, a basket and an oil lamp, chipping away at the rockface, and then bringing the ore out. For those wishing to find other similar mines, there are good examples in altered and mineralized pillow lavas on the south side of the E802 between Mandria and Pera Pedi (36 486845 E, 38 58053 N), and in steeply dipping sheeted dykes on the hillside to the north of the E703 in Asprogia (36 464606 E, 38 64533 N).

Stop 7.3 Sulphide deposits, ancient slag and modern copper production at the Skouriotissa mine Skouriotissa is an impressive and important location, but access is not guaranteed because of fluctuations in mining activity that sometimes result in closure of the mine. Those wishing to visit must first check that this is possible and then book a tour or participate in an open day run by the mine operator at the time.* The information below provides only an overview of the geology and mining activities at Skouriotissa. For details on the regional setting of the copper deposits exploited, see stops 3.5 and 3.6, and Figure 3.3.

The mine is most conveniently reached from Linou, which lies along the E908 about 3 km west of its junction with the B9 Lefkosia–Troodos road. At Linou, where the spoil and leach heaps of the mine loom up to the north, turn north and travel through Katydata to the outer limit of Skouriotissa village, where the mine entrance is off to the right (36 489833 E, 38 82890 N). A little beyond the mine entrance, opposite a United Nations camp, there is an excellent exposure of ancient slag. Beware of heavy vehicles on approaching and leaving the mine.

The mineralization at Skouriotissa occurs in the uppermost part of the volcanic sequence along the axis of the Solea graben. This is unlike all of the other sulphide deposits in the Troodos ophiolite, which occur lower in the volcanic sequence. In recent decades, mining has focused on three orebodies. The Foucassa orebody was a lens of massive pyritic ore that was 670 m long, 210 m wide and 150 m deep, and was capped by ochre and umber, and overlain by white marl (see Fig. 3.16a). In contrast, the Phoenix orebody comprised highly oxidized, disseminated, vein-type mineralization that may have been part of a stockwork zone or may have formed by supergene enrichment. The latter process was probably responsible for the formation of the Three Hills orebody, but it is of lower grade than the Phoenix body and is unrelated to it.

Skouriotissa is one of the oldest (at least 2760 BC) and most important copper-producing areas in Cyprus, as is confirmed by the presence of half of the ancient slag of the entire island in the local area. Even the name of the local church (Our Lady of the Slag Heaps) reflects this importance. The

---

* In 2009 the operator was Hellenic Copper Mines Ltd (e-mail info@hcm.com.cy; telephone +357 22 933312; fax +357 22 933311; website www.hcm.com.cy/indexenglish.html).

ancients not only mined surface exposures of ore, but they also developed extensive underground workings. The last major phase of mining at Skouriotissa began in 1921. Early activity was mainly underground, but between 1963 and 1973 open-pit mining took place at Foucassa until most of the massive ore had been extracted, which amounted to nearly 7 million tonnes in total at an average of about 2.5 per cent copper. In 1972 attention turned to the lower-grade Phoenix orebody, which also contained gold and silver. The oxidized nature of the ore meant that copper could be effectively extracted by acid leaching of crushed ore and, later, of exposed rockfaces. Copper was recovered through the regular addition of scrap iron to the leachate.

Since 1996, Hellenic Copper Mines Ltd has been using a highly efficient hydrometallurgical process to extract copper from the low-grade ores of Phoenix, Three Hills and elsewhere, and mineralized spoil from the nearby Apliki mine. Low-grade rock (< 0.5% copper) is dumped in heaps without crushing, whereas higher-grade rock (0.5–2.0% copper) is first crushed to a diameter of 75 mm before being placed in layered heaps (Fig. 7.4a). Dilute sulphuric acid is sprinkled or dripped onto these rock piles to leach copper; drippers are preferred because they reduce evaporation. The process is particularly effective if the acidified piles are warm and well oxygenated to encourage the growth of the bacterium *Thiobacillus ferrooxidans*, which catalyzes the breakdown of sulphide minerals and liberation of copper. Heaps are leached for at least 36 months, and this removes about 75 per cent of the copper. Drainage pipes at the base of the piles collect the leachate and transport it to storage ponds (Fig. 7.4b), from where it is repeatedly recycled through the heaps until its copper content reaches an acceptable level, at which point it is pumped to holding tanks. From here the leachate is mixed with an organic reagent that removes the copper. Following separation of the two liquids, the depleted leachate is returned for re-use in the leaching process, and the organic reagent is linked to an electrolyte stream, to which it releases its copper. Copper is removed from the electrolyte by electrolysis, using lead anodes and stainless-steel cathodes, and is deposited on the cathodes (Fig. 7.4c). This copper is extremely pure and is exported to Greece to be made into thin high-quality wires. It was also used in the manufacture of the bronze medals awarded at the Athens Olympic Games in 2004.

Water is essential to mining operations at Skouriotissa because of the leaching process. Water has to be carefully managed in order for there to be enough during the long summers. Consequently, the open pits double as reservoirs to store both rainwater and water pumped up from nearby rivers during the winter. When the Foucassa deposit was worked out, the deep pit was lined with bentonite and now stores water all year. The

Figure 7.4 Hydrometallurgical extraction of copper at the Skouriotissa mine (stop 7.3): **(a)** leach heaps; **(b)** a storage pond at the base of leach heaps; **(c)** sheets of copper plating stainless-steel cathodes.

Phoenix pit is used as a reservoir to a lesser extent because minerals remain to be exploited.

The hydrometallurgical operation at Skouriotissa is designed to have minimal impact on the environment. The recycling of leachate keeps water consumption to a minimum, and the self-contained nature of the process means that it is unlikely that solvent and leachate will escape into the local environment around the mine. Should solvent or leachate find a way through the impermeable high-density polyethylene membrane at the base of the leach heaps, it should be neutralized by an underlying layer of limestone. There is in fact an element of environmental restoration following leaching, because, once the leached rock piles are unable to yield further copper, they are neutralized using limestone and then covered with fertile soil and planted with trees. In this way the scars of mining are somewhat reduced.

Stop 7.4 Ancient slag, pillow lava, limestone and gypsum quarries, and sulphidic tailings north of Mitsero The village of Mitsero lies about 25 km southwest of the centre of Lefkosia. Its surrounding area records a long history of resource extraction, processing and waste disposal that dates back to at least Roman times, and this is examined in a 4 km-long excursion along the F922 and part of the F908 north of Mitsero. The village sits on the volcanic sequence of the Troodos ophiolite, which is mineralized in places and has been mined near by at the Agrokipia and Kokkinopezoula mines, but the hills to the north are dominated by carbonate rocks of the Pakhna Formation. The excursion begins by looking at ancient slag, concludes by examining large tailings piles and, in between, provides an overview of large-scale quarrying.

Mitsero lies immediately north of the E905, from which it should be entered by taking the easternmost (closest to Agrokipia) turnoff. Follow the road down into the village and turn right onto the F922 towards Kato Moni. Shortly after leaving the village, two shallow cuttings are encountered on the right side of the road; these expose ancient slag (36 511744 E, 38 78011 N).

These slags are vitreous and black (see Fig. 7.2b), which suggests that they date back to Roman or later times. The dip of the slags demonstrates that the molten waste from smelting the ore was poured onto growing piles of slag. Some idea of the efficiency of the smelting process may be obtained by examining the slag for copper staining, which suggests that not all of the metal was transferred to the smelt. These slag heaps, like most of those in Cyprus, have been seriously disturbed because of their use for road metal and, to a lesser extent, as a source of silica and iron used in the manufacture of cement. Fortunately, their historical importance has been recognized and many of the remaining slags are protected antiquities, as they are here; the sign posted on top of the piles translates as "Department of Antiquities. Place of Historic Monument. It is forbidden to remove material. If you do, you could be taken to court." Please respect this warning and do not remove anything from the site.

Continue on the F922 and, shortly after crossing a small bridge, pull off and park on the left at the start of a track. Walk a short distance up this track to the second outcrop on the right, which exposes superb pillow lavas that exhibit settling of olivine crystals and radial cooling joints (36 511330 E, 38 78450 N). From here, the view east across the valley reveals the remains of the abandoned Hellenic Mining ore-dressing plant, which was operational until the late twentieth century. This plant received ore from the Agrokipia and Kokkinopezoula mines, and separated the ore minerals from the worthless rock. The resulting ore concentrate was then transported to the loading station of Karavostasi, northwest of the Skouri-otissa mine, for shipping abroad. The plant also used cyanide leaching to extract gold and silver from the ore. The tailings generated by the dressing plant are distributed over the surrounding area, and one of the main dumps is examined at the last locality.

Return to the F922 and follow it up hill to the north, through a very thin horizon of the Lefkara Formation and then into the Pakhna Formation. Caution must be exercised on this stretch of road, because it is used by heavy lorries and it may not be suitable for large coaches. On top of the hill, at the end of a relatively straight and flat piece of road shortly before the entrance to a quarry where the road bends sharply to the left and descends, the road is wide enough to park on (36 511608 E, 38 79986 N).

The parking place provides a good overview. To the southwest, west and northeast are large quarries where Koronia reef limestone of the Pakhna Formation and, to a much lesser extent (and mainly in the past), gypsum of the Kalavasos Formation are being extracted. The scale of the operations is impressive and much dust is created during the quarrying and processing of the limestone, which is used primarily for lime and crushed stone. To the north and east are eroding piles of yellow, orange, brown and grey sulphide-bearing tailings that rest on carbonate rocks of the Pakhna Formation. These are investigated in detail at the next location.

Cautiously follow the road as it descends, staying aware of its uneven surface and the many heavy vehicles, and then veer sharply around to the right at the crude crossroads near the bottom of the hill and head east. This new road is the F908 to Agios Ioannis and it crosses badland topography resulting from the instability of the underlying marl-rich lithologies of the Nicosia Formation. After 1.5 km on the F908, having passed a huge pile of gullied tailings and a roadcut, park on a small dirt track off to the right

Figure 7.5 Sulphidic tailings lying on top of carbonate rocks of the Pakhna Formation (stop 7.4). The hill is covered by a flat-top tailings pile, which is eroding into a series of tailings ponds, one of which occupies the foreground. The absence of water in the pond is attributable to a breach in the tailings dam, and the escaping water has cut down into the tailings.

(36511947 E, 3881143 N). This parking place is not suitable for coaches, but a coach party may be dropped off here and picked up later.

Walk in a southeasterly direction along the track, keeping the tailings ponds on the right. The track follows the tops of crude tailings dams that periodically impound acidic drainage water. Channels in the tailings and a breach in the dam of the second pond reveal stratified sediment (Fig. 7.5), with dark-grey pyrite horizons sandwiched between orange to yellow laminated horizons. The latter probably contain large quantities of iron oxides (e.g. goethite) and iron oxyhydroxysulphates (e.g. jarosite). The tailing-retention system is not working effectively, as is demonstrated by the breach in the dam and the occurrences of tailings down stream. Farther along the track there are other tailings ponds, and a short distance up slope there is a narrow concrete dam with drainage pipes, which no longer functions because of sedimentation and disrepair. The tailings were piled on carbonate rocks of the Pakhna Formation, presumably in an attempt to reduce the acidity of waters draining from the piles.

Industrial and construction materials

The distinction between industrial minerals and construction materials can be unclear, because sometimes the same commodity is listed under both of these classes of resource. Industrial minerals are valued for their specific physical or chemical properties, rather than for any constituent metals, and they are commonly monomineralic sedimentary rocks, such as carbonates, clays and evaporites. Cyprus hosts many deposits of industrial minerals, which include: chalk for whiting; montmorillonite clay (bentonite) for drilling mud, impermeable liners and barriers, cat litter and mineral filler; gypsum for fertilizer and mineral filler; and umber for mineral pigment. Other deposits of note are asbestos, magnesite, halite, celestite and quartz (silica sandstone).

In contrast to industrial minerals, which (like metallic minerals) tend to occur in localized concentrations, construction materials are both abundant and widely distributed and, therefore, have low intrinsic value. Once again, Cyprus is fortunate in its abundance of construction materials, such as: limestone, calcarenite and marble for building and armour stone; limestone, dolerite and sand for aggregate; limestone, chalk, gypsum and clay for cement; sand for mortar and concrete; clay for bricks; and gypsum for plaster. Alluvial sands and gravels are potentially attractive construction materials, but it is illegal to extract them, because they constitute important aquifers. Because of their bulk and density, construction materials are costly to transport and ideally they are processed (cut, polished, crushed or roasted to increase their worth significantly) and used close to their point of extraction. A good example of this is provided by the cement industry on the south coast of Cyprus east of Lemesos, where the cement works at Vasiliko and Moni utilize the nearby deposits of chalk, marl, clay and gypsum (stop 6.6).

In addition to the stops described below, a variety of industrial minerals and construction materials may be examined at many past and present extraction sites: bentonite (stop 5.6), chrysotile asbestos (stops 2.4, 7.18), halite (stop 6.17), silica sandstone (stop 6.2), umber (stop 3.15), gypsum (stops 6.12, 6.13, 7.4), limestone (stop 7.4) and blocks of building stone (stops 5.9, 7.15, 7.16). Excellent examples of the ancient uses of local and imported construction materials can be seen at stops 7.9, 7.15 and 7.16.

Stop 7.5 Building stone and stone implements at the Choirokoitia Neolithic settlement From the A1 motorway take exit 14, turn onto the F112 and follow signs for Choirokoitia Neolithic settlement. Park in one of the areas near the site entrance (36 531555 E, 38 50503 N); there is a small fee to enter the site. This stop may be conveniently combined with stop 6.10 near by.

The site was inhabited during two periods in the Neolithic: a pottery-devoid (aceramic) period from about 7000 to 5800 BC and a pottery-making (ceramic) period from about 5000 to 3900 BC. Such is its significance that it is a UNESCO World Heritage Site, designated such because of its excellent state of preservation, its importance for human settlement studies in the Mediterranean region, and the important role of Cyprus in the expansion of civilization from the Near East to Europe.

The site consists of a reconstructed village and an excavated area on the slope of a hill above the western bank of the Maroni River. The village may have been sited here because of the river, perennial springs near by, arable land and the defendable position. Taken together, the circular reconstructed buildings and excavated remains demonstrate that walls were made of mudbrick (a mixture of sun-dried soil and straw) or pise (a type of mud wall) above stones (gabbro, microgabbro and dolerite of the Troodos ophiolite, calcarenites and cherts of the Lefkara Formation and reef limestones of the Pakhna Formation), which were cemented with mud mortar (Fig. 7.6a). The rounded ophiolitic and sedimentary stones were probably collected from the nearby river, as can be confirmed from the shapes and compositions of boulders in the present-day river bed. The angular Pakhna rocks, by contrast, may have been quarried from outcrops within the village.

Locally derived chert, flint, limestone, dolerite, microgabbro and animal bones were used to make tools, grinders, and pounding and crushing implements, some of which are visible around the site (Fig. 7.6b). Jewellery, pendants and beads were made from picrolite that probably brought to the village from the bed of the Kouris River west of Lemesos, where it is found in abundance in sediment derived from erosion of Troodos serpentinite. Obsidian has also been found and was used for making small retouched blades (sharp stone fragments modified for a particular use). This rock type does not occur in Cyprus and was brought to the site from south-central Anatolia, probably via the southern coast of Turkey or through northern Syria.

Just 6 km southwest of Choirokoitia is another Neolithic site, Kalavasos–Tenta (36 527773 E, 38 45594 N). These two villages are the same age and exhibit many similarities, but their relationship to one another is unclear.

Stop 7.6 Building stone, stone implements and water resources at the Lempa Chalcolithic site north of Pafos It is best to reach the Lempa Chalcolithic village from the E701 Tomb of the Kings road between Pafos and Coral Bay. Along this road, 6 km from its origin in Pafos, there is a brown sign with yellow lettering indicating that the archaeological site (and the Lempa pottery) is along a concrete road leading inland. There is no prior

Figure 7.6 Archaeological finds at Choirokoitia Neolithic settlement (stop 7.5): **(a)** stone foundations of circular buildings; **(b)** microgabbro quern and rubbing stone (the coin is 28 mm across).

warning for this turning, so aim to turn inland almost opposite the Helios Bay hotel apartments. About 800 m along the concrete road there is a parking area in front of the site (36 445617 E, 38 52500 N).

An experimental village, with circular buildings made of mud walls on stone foundations with timber and earth roofs, has been erected on the site where a community existed on the seaward edge of modern Lempa from about 3500–2400 BC. Like many other ancient settlements in Cyprus, the village was sited near a perennial spring, here on the 50 m marine terrace. The spring lies at the contact between calcarenite and underlying impermeable Kannaviou clay, which acts as an aquiclude. Groundwater seeps through the calcarenite, encounters the clay and runs along the clay/calcarenite contact until it emerges as springs.

The site exemplifies many features characteristic of a Cypriot Chalcolithic village, and finds from the site are now exhibited in the Pafos museum and the Cyprus museum in Lefkosia. The occupants of Lempa probably obtained their raw materials from nearby stream beds and beaches, where a wide variety of rock types occur as smooth rounded pebbles and cobbles. Rocks derived from the Troodos ophiolite were used extensively: basalt was made into small thin tools, such as chisels, adzes and polishers; pyroxene andesite was used for fine axes; dolerite and gabbro for large axes, hammerstones, grinders and pestles; and serpentinite for small pebble polishers. Dense grey and crystalline Mamonia limestone was turned into pounders, and dense and silicified Lefkara chalk was used for making bowls, hammerstones, grinders, pestles and rubbing stones. Light-yellow calcareous sandstone and porous calcarenite, derived from raised-beach deposits along the Pafos lowlands, were used for pestles, rubbing stones and cupped stones.

Stop 7.7 The sand quarry near Agios Sozomenos north of Dali The quarry lies about 4 km north of Dali and is reached most easily by heading east towards Agios Sozomenos on the E120 after leaving the A1 motorway at junction 6, 14 km south of the centre of Lefkosia. Several kilometres from the motorway, there is a T-junction. Turn right here and drive southeast towards Potamia for just over 1 km to reach a crossroads, at which point turn left towards Geri and head up hill, keeping right where the road forks as it passes through a large farm. The entrance to the quarry is at the top of the hill (36 538856 E, 38 79554 N). The quarry is owned and operated by Latouros Quarries Ltd, who are very willing to accept unannounced visitors, but it is polite to make contact prior to a visit.[*]

The quarry is extracting very friable shelly calcarenite of the Nicosia Formation for use as sand aggregate in the production of concrete, bricks and mortar. The calcarenite is crushed, screened and washed on site, with the screening removing shell fragments. Washing involves a very modern water-recycling system that enables re-use of much of the 1000 m^3 per hour of water required to clean the sand of very fine material, and this system

---

[*] E-mail latouros@latouros.com; telephone +357 24 030020; fax +357 24 030019; website www.latouros.com.

also minimizes discharge of waste slurry. The absence of marl and clay in the slurry indicates that these highly undesirable components are not present in the original calcarenite, making it a high-quality aggregate. The screened and washed sand has a fine grainsize and is well rounded, and this makes for good workability in the production of concrete, which reduces the amount of water needed, thereby giving the concrete more strength.

Stop 7.8 The magnesite mine north of Smigies on the Akamas Peninsula A magnesite mine is located north of Smigies picnic site on the Akamas Peninsula to the west of Polis. It should be visited on foot, following the nature trail from Smigies, or in a four-wheel-drive vehicle, as parts of the gravel roads and tracks are very rough. Smigies lies to the west of Neo Chorio, which is reached from Prodromi by following the E713 west along the coast, through Lakki, and then turning inland onto the F735. Drive carefully through the narrow streets of Neo Chorio, keeping the main church on the right, and then continue due east to Smigies picnic site, passing the small church of Agios Minas on the way. If proceeding by vehicle, travel beyond the picnic site to a T-junction and turn right here to head due north towards Fontana Amorosa. This road runs along the top of the ridge of the northern part of the Akamas, which is mostly made up of a wide range of rocks of the Troodos ophiolite, and it affords spectacular views to the west. After travelling about 1.7 km from the T-junction, and just before two valleys running down and away on both sides of the dirt road, turn right onto a track running north-northeast, parallel to a valley on the left and a ridge on the right. About 500 m from the turnoff, below a very prominent outcrop of Koronia reef limestone, is the entrance to the mine (36 438683 E, 38 77819 N).

The mine is a cutting into serpentinite on the uphill side of the track, and spoil lies on both sides of the track. The narrow entrance opens into an oval pit in which there are tunnels into the pit wall. The main rock type is a highly sheared and shattered serpentinite, and this is cut by an interconnected array of mostly steeply dipping veins of white magnesite up to 30 cm wide. The magnesite is clearly younger than the serpentinite and has a botryoidal texture that looks just like a head of cauliflower. The deposit lies close to the overlying carbonate rocks of the Pakhna Formation, which crop out a little farther up hill. This relationship is exhibited by most, if not all, of the major concentrations of magnesite in the area, suggesting that they were formed during alteration of serpentinite when it reacted with downflowing carbonate-rich water.

The proportion of recoverable magnesite to country rock appears to have been quite low. Despite this, mining took place sporadically during the twentieth century. The separated magnesite was roasted on site to produce magnesium oxide that was then shipped abroad, mainly to be used to line furnaces, particularly those used in the steel industry. The roasting kiln is near by.

Hydrogeology and water resources

The water situation

Water is by far the most important natural resource on Cyprus, and even the ancients constructed elaborate aqueducts, pipes and storage systems to bring water into their towns and cities. Today, water is critical for the two most important industries in Cyprus – agriculture and tourism – but the island suffers from an acute shortage of this vital resource. The shortfall between limited supply and excessive demand is periodically worsened by severe droughts, the most recent of which occurred in the late 1970s, early 1980s and since the late 1990s. The high demand for water has led to over-abstraction of groundwater, the construction of desalination plants and many dams, and the implementation of irrigation and water-conveyor schemes. Even these cannot meet current demand and there are plans to improve water management by developing efficient irrigation methods and a more highly integrated system of water capture, storage and supply.

Cyprus has a Mediterranean climate; its total precipitation averages 500 mm per year (total volume 4600 million m^3), but varies from as low as 300 mm in the central Mesaoria Plain to over 1000 mm on the highest points of the Troodos massif (see Fig. 1.4). The heavily dissected topography of the Troodos massif clearly demonstrates that all of the major rivers in Cyprus originate in this mountainous area (see Fig. 1.2). These rivers account for most of the 13 per cent of the total precipitation that ends up as surface flow. The remaining water from precipitation either is lost through evaporation and evapotranspiration from non-cultivated areas (77%) or recharges groundwater supply (10%). Two-thirds of the groundwater supply is recovered through pumping, wells and springs, but the remainder is lost to the sea. Of the surface flow, most is still lost to the sea, despite the extensive damming of many of the main rivers to the point that there are no longer any perennial rivers flowing to the sea as there were 100 years ago. Only 3 per cent of the total annual precipitation is captured and stored in reservoirs, but just over half of this amount is actually used; the rest is probably lost predominantly through evaporation.

Groundwater

Aquifers are the main source of water in Cyprus, and their distribution clearly reflects the geology of the island, with most contained within the circum-Troodos sedimentary succession (compare Figs 1.2 and 7.7). The major aquifers in Cyprus can be subdivided into three groups based on lithology and, to a lesser extent, structure. The first-class aquifers include the western and southeastern parts of the Mesaoria Plain and the Akrotiri Peninsula, and comprise unconsolidated alluvial sands and gravels, and porous sandstone and calcarenite. These contribute about half of the groundwater supply, despite the fact that they reside below less than 20 per cent of the surface area of Cyprus. The second-class aquifers occupy a similar subsurface area, but supply only 10 per cent of the total groundwater. They occur in fractured and karstic chalk, limestone, dolomite and gypsum deposits, where water is stored and transported mainly through solution channels. The remaining 60 per cent of the subsurface area of the island is occupied by aquicludes of impervious rocks. However, there are aquifers in semi-permeable sedimentary beds and highly fractured igneous and metamorphic rocks; for example, in the Troodos ophiolite, groundwater is stored in fractured gabbros and serpentinites, and emanates from relatively abundant perennial springs in these rocks (e.g. stop 2.4). In fact, many of the Troodos hill villages are located along the gabbro/dyke contact because it is the site of springs, and weathering of gabbro produces sandy sediment, which makes a good shallow aquifer (e.g. stop 2.9).

Water supply, quality and management

Domestic supply accounts for about 20 per cent of the water used in Cyprus, and tourism a mere 2 per cent, in stark contrast to the 78 per cent consumed by irrigation for agriculture. Aquifers are under huge stress, and water consumption is currently estimated to exceed the safe yield from groundwater resources by about 43 million m^3 per year (island wide), an amount equivalent to about 1 per cent of the total annual precipitation. Investigations have shown that there are no large untapped groundwater resources on the island, which means that this annual depletion is hugely significant. Already sea water has intruded into overpumped coastal aquifers, and the water table in mountainous areas has been drawn down significantly. Various actions have now been taken to combat this problem of over-abstraction of groundwater: an efficient drip system and treated wastewater are now used for irrigation; desalinized water is being used for domestic purposes, but it is very expensive and energy-intensive to produce; land consolidation has increased the efficiency of water use; and methods to store water underground are being considered.

With the over-exploitation of groundwater, surface reservoirs remain

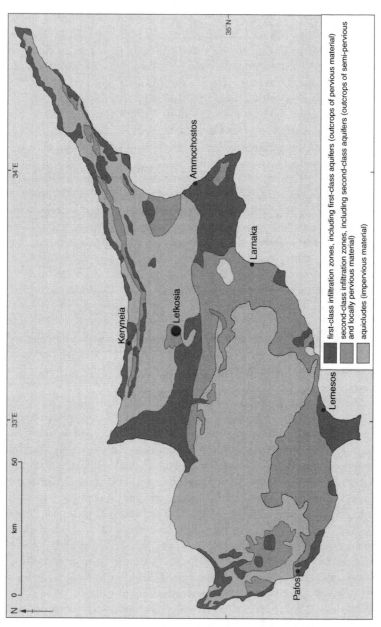

Figure 7.7 Distribution of aquifers and aquicludes in Cyprus (modified from Cyprus Geological Survey Department 1970).

important. Hence, many rivers have been dammed, especially in response to the government's water policy since independence in 1960, which is captured in the slogan "not a drop of water to the sea". In 1960, reservoir capacity stood at 6 million m^3, but it is now at 300 million m^3 and there are plans to add a further 100 million m^3 of storage despite most of the exploitable rivers having already been dammed. This huge rise in storage capacity has resulted mainly from the construction of many large earthfill or rockfill embankment dams in broad flat-bottom valleys within the circum-Troodos sedimentary succession and the upper sequences of the Troodos ophiolite on the periphery of the Troodos massif. In fact, according to the International Commission on Large Dams, Cyprus now ranks first in Europe for the number of large dams per unit area.

Two of the most notable water-management schemes that have led to this considerable rise in reservoir storage are the Pafos Irrigation Project and the Southern Conveyor Project. Both of these are large-scale infrastructure schemes put in place to transfer water to water-poor agricultural areas in southwest and southeast Cyprus. The Pafos Irrigation Project was developed between 1976 and 1982; it uses a series of canals, pumping stations, irrigation networks and boreholes to transfer water, mainly from the Asprokremmos reservoir on the Xeros Potamos, to the coastal plain in the Pafos area. The Southern Conveyor Project is the largest water development project in Cyprus; it was developed between 1984 and the late 1990s. The project area extends a considerable distance along the south of the island, from the Diarizos River in the west to the main area for producing potatoes – Kokkinokhoria in the southeast. Surplus water from the Kouris reservoir and other large reservoirs farther west is conveyed by pipeline to the low-rainfall highly cultivated areas in the southeast.

One key aspect of water management is the maintenance of water quality, especially where water resources are limited and under extreme stress. Recent regional surveys in Cyprus have shown that waters contain elevated levels of magnesium, chromium and boron when compared with drinking-water standards set by the World Health Organization. However, these enrichments may be attributed to reaction of water with rocks of the Troodos ophiolite that have relatively high levels of these elements and which are the source of much of the water. These same surveys have also shown that surface waters and, particularly, groundwaters have been adversely affected by contamination from agriculture (mainly nitrates), seawater intrusion (salt), acid mine drainage from abandoned sulphide mines (sulphate, metals and arsenic), and asbestos from the mining and processing of this mineral. The risk posed by these contaminants is assessed on the basis of water use. For example, some farmers use acid mine water on their fields, with no obvious deleterious effect, whereas nitrates that accumulate in silt residues in reservoirs used to supply drinking water are a cause for concern.

In addition to the stops described below, there are many others that complement them. The importance of sources of water for locating settlements is highlighted by stops 2.9, 6.8, 7.5 and 7.6, all of which demonstrate the importance of springs, with further examples of these sources of water at stops 2.4 and 5.9. There is an excellent example of ancient water supply and storage at stop 7.15, and an earthfill embankment dam can be seen at stop 5.8. An appreciation of efficient recycling of water during mineral and rock processing can be gained at stops 7.3 and 7.7, whereas contamination created by such processing and by waste disposal is considered at several places: acid mine drainage at stops 3.15 and 7.4, sediment contamination at stop 7.18, and landfill discharge at stops 7.13 and 7.14.

Stop 7.9 The ancient city, water resources, sea-level change and Pakhna and Quaternary rocks along the Amathous circuit east of Lemesos The archaeological site of Amathous lies at the eastern end of the sprawling coastal strip of hotels and restaurants east of Lemesos, and signs clearly indicate the site entrance on the landward side of the B1 (36 513054 E, 38 40951 N). The appropriate section of the B1 may be reached from the A1 motorway via junctions 21 (to the east of the site) or 22 (to the west of the site). There is parking at the site entrance and a small entrance fee. The details given here describe a round-trip of several kilometres, which starts and ends at the archaeological site, and includes examination of Pakhna and Quaternary rocks. Done completely, the circuit can take a full day.

Although there were Neolithic settlements in the hills around Amathous, the actual site is believed to have been uninhabited until the Late Bronze Age (around 1100 BC). Growth of the settlement was favoured by the natural setting: the surrounding hills were used for defence and agriculture, trees and water were abundant, and the sea offered a means of transport and anchorage. Local limestone quarries provided building stone for the port and city monuments, and mining of copper at Kalavasos, Parekklisia and north of Armenochori, as well as in the Limassol Forest, brought great wealth to the city.

Amathous was one of the oldest city kingdoms in Cyprus. By the eighth century BC, the site was under the control of the Phoenicians, who used it as a base for trading. Between the eighth and sixth centuries BC, Phoenicians, Hittites, local Cypriots, Greeks and Egyptians ruled Cyprus, and Amathous was influenced in turn by each of these cultures, becoming a focal point for trade. During the Greco-Persian wars in the fifth and fourth centuries BC, Amathusians sided with the Persians, remaining in the process one of the independent kingdoms in Cyprus headed by a local king.

Following the defeat of the Persians by Alexander the Great, King Androkles of Amathous attempted to maintain power, but failed when Amathous became a democracy in 312–311 BC. The city was then under the domination of the Ptolemies, Romans and Byzantines until its decline in the Arab raids of the seventh century AD.

Perhaps the most impressive part of the site is the agora (marketplace), which lies at the end of the road from the site entrance. An information board at the southeastern end of the agora provides a comprehensive orientation. The water-supply infrastructure and baths are particularly noteworthy. At the northwestern end are a nymphaeum (monumental fountain) associated with a reservoir, and a vaulted reservoir and cistern, which were all tapped by a series of limestone and terracotta conduits that supplied water to the Roman baths, a centrally positioned Roman public fountain and, in earlier times, Greek baths (Fig. 7.8). Some fine columns remain on the agora site. Those made of limestone clearly come from the Pakhna Formation, but the granite, white-marble and well foliated grey-marble columns must have been imported, probably from Turkey, as no such rocks occur in Cyprus. The importation of such exotic rocks confirms the wealth that this ancient city once possessed.

The water-supply network to Amathous is extensive and superbly integrated, as it reached as far north as Armenochori and Parekklisia, and consisted of a complex linked system of springs, reservoirs, conduits,

Figure 7.8 Circular limestone water conduit (centre foreground) leading from the fountain reservoir (behind column) adjacent to the vaulted reservoir (with arch) in the agora of the ancient city of Amathous (stop 7.9).

channels and aqueducts. About 500 m due north of the site entrance is the north wall, which was built as a defence in the low-lying area between adjacent hills. Looking north from this wall, it is possible to see the rubble foundations of a once-arcaded section of the main aqueduct, which carried water from springs in the hills around Armenochori. The north wall here supported a feeder system for water entering the city, and an offshoot from this system probably supplied water to the fountain reservoir in the agora. Much of the infrastructure for the capture, storage and transport of water seen today dates to the Hellenistic and Roman periods.

Immediately south of the agora is the harbour. The inner harbour lay in what is now the sandy depression between the ticket office at the site entrance, the agora and the B1 coast road. This once-natural cove appears to have been abandoned in the late fourth and early third centuries BC, because the outer harbour silted up. The outer harbour was constructed late in the fourth century BC and lies to the south of the inner harbour, and from the coastal footpath south of the B1 may be seen lying submerged immediately beyond the present-day coastline. By the first century AD a beach had developed well out into the outer harbour as a result of elevation of the land and deposition of sand and silt by both water and wind. The migration of the shoreline back inland since then is a sign of relative changes in sea level during the past two millennia. Further evidence of this sea-level change may be seen about 150 m farther east along the coast from the harbour, where remains of the fifth-century AD southeast basilica are partially submerged. These changes in relative sea level reflect the rising and sinking of the land in this seismically active region, rather than global changes in levels of the oceans.

About 400 m west of the site entrance is a small broad valley with excellent exposure of the Pakhna Formation and interesting archaeology. It is reached by following the path from the entrance on the landward side of the B1. The path passes the southwest basilica and then remnants of the southwest wall and west gate, which date from Archaic and Classical times, just before it goes under the B1. At this point, the wall, gate and rockface mark the edge of a relatively flat area of the valley bottom that contains remains of land snails and abundant fragments of ancient pottery (36 512718 E, 38 40883 N), which must not be removed. The rockface may once have been a seacliff, but it has also been quarried. At its base there is a manmade water channel that has been cut into the rock and this probably tapped off water from springs farther up the valley. While exploring, be aware of the potentially dangerous excavations at the foot of the cliff.

The Pakhna rocks in the quarried cliff face are well bedded carbonates, ranging from calcarenite to sandy marl. They exhibit a variety of sedimentary features, and graded bedding, convolute stratification, slump folds, flame structures, load casts, channel fills and intraclasts may all be seen (see Fig. 6.7b). Many of these features are characteristic of loading and slumping of soft sediment, perhaps accompanied by de-watering, in a high-energy shallow-water shelf-slope environment. Further evidence of soft-sediment deformation can be found on the western side of the valley, west of the Moulin Rouge club, on the sharp bend in the road that comes inland from the B1 (36 512599 E, 38 40974 N).

Now take the path under the B1 and examine the impressive remains of the Hellenistic southwest wall on the coast path. From here the coastal boardwalk leads back to the main archaeological site, but it is best to head about 400 m west to Loures Beach, below the Avenida Beach hotel, in order to undertake a full transect of the clearly exposed geology along the coast between beach and archaeological site. At Loures Beach (36 512336 E, 38 40759 N), the narrow zone of black ophiolitic sand gives way to dark ophiolitic beachrock that dips very gently seawards (see Fig. 6.14c). The key feature in the beachrock is the fragments of orange and reddish-brown pottery, which limit the oldest age of the rock to within historical times, probably the past 2000 years, and demonstrate how quickly sedimentary rocks can form. The cement of the beachrock is carbonate, seen coating many of the clasts, and this crystallized during repeated and rapid wetting and drying in the zone of wave splash.

About 100 m east (36 512462 E, 38 40787 N), the beachrock gives way to conglomerate overlying Pakhna limestone along an uneven unconformity. The conglomerate, dominated by ophiolitic pebbles, exhibits alternations of coarser and finer bands, some of which partially overlap. These features, and the presence of shell fragments, suggest that the conglomerate is of Pleistocene age and that it originated as part of a pebbly beach or shallow-marine debris flow. The abundance of ophiolitic pebbles, with the exception of rock types from the mantle sequence, implies that the main sediment source was the crustal section of the Troodos ophiolite. The underlying bioclastic limestone has a highly eroded surface, much like that seen at sea level farther east, and it has been bored by marine organisms. This eroded and bored surface was once below sea level, but is now above it, again demonstrating relative sea-level changes along this section of shoreline.

Eastwards, the coastal path soon becomes a boardwalk over wavecut bedding surfaces of Pakhna calcarenite. These surfaces are highly grooved, probably as a consequence of wave and pebble action and dissolution, and there are also grooved and bored terraces, some having patches of conglomerate containing abundant ophiolitic rock fragments. Infilled channels and fissures of Pakhna age occur in places, as do flame structures. The bedding surfaces of some of the coarser beds reveal corals, echinoids, shells and shell fragments; the trace fossils *Chondrites* and *Thalassinoides* are present at several horizons. Many of the larger shells are in a concave-up orientation, suggesting that they settled in suspension. All of these sedimentary features and fossils are very well exposed at 36 512914 E, 38 40839 N.

Stop 7.10 Water storage, slope instability and Pakhna Formation at the Kouris dam The Kouris dam lies northwest of Lemesos and is reached from the A6 motorway by taking the exit for Ypsonas and Kouris dam, and then turning right onto the F816 and following this road northwestwards for several kilometres. At the brow of a ridge where the dam first comes into view, park on the left, opposite an old stone building with blue shutters (36 493207 E, 38 41229 N). From here there is an excellent view up the Kouris Valley of the imposing dam and the hills and cliffs made of rocks of the Pakhna Formation.

The Kouris dam was completed in 1988, and at 550 m long and 110 m high it is the largest in Cyprus (Fig. 7.9a). The dam is a zoned earthfill embankment structure with zones increasing in permeability from the centre outwards. The central impermeable clay core reduces seepage, whereas the weight of the heavy outer zones provides most of the stability. This type of structure is designed to accommodate deformation, such as that caused by settlement and earthquakes. The vast size of the base of the dam means that it imposes relatively low stress on its foundations when compared to gravity dams (see stop 7.11). The dam is built on a foundation of well bedded chalks, calcarenites and marls of the Pakhna Formation, which dip at less than 10° down stream. Beds of coarse calcarenite present a problem because they are highly permeable and they provide horizons for leakage. Despite the presence of an impermeable barrier (a cut-off) built into the rocks at the base and sides of the dam, it has been estimated that at full capacity the reservoir would lose up to 1.5 per cent of its water every year through these permeable calcarenites.

To take a closer look at the dam, the reservoir and the valley sides, drive farther northwards along the F816 and park on the eastern abutment of the dam, in the layby in front of the control building. There is ample room here to park and turn a coach. A walk along the dam crest will reveal the village of Alassa at the northern end of the reservoir, which was relocated from its original site (now under water). The reservoir (maximum storage capacity 115 million m³) is the largest in Cyprus and it provides the main store of surface water for the Southern Conveyor Project, drinking water for the Lemesos area, and water for irrigation of crops. Since its completion, severe droughts have prevented the reservoir from reaching full capacity. Such was the draw-down of the reservoir level in the late 1990s

Figure 7.9 Examples of the two most common types of dam in Cyprus: **(a)** the 550 m-long Kouris earthfill embankment dam spanning the broad Kouris Valley (stop 7.10); **(b)** the 132 m-long Palaichori–Kampi concrete gravity dam in the V-shape Kampi Valley (stop 7.11 viewed from Apliki). (Image (a) kindly supplied by the Cyprus Water Development Department)

and early 2000s that the lowermost part of the upstream side of the dam was exposed. Linked to these very low water levels were incidences in Lemesos of drinking water enriched in bacterially produced hydrogen sulphide, which was caused by low levels of oxygen in the reservoir water.

A walk along the dam crest will also provide evidence of settling and differential movement within the dam, which is a common feature of structures of this type and size. The pavement half-way along the crest road is buckled and some of the paving slabs have cracked. At the western end, where the embankment meets the spillway, the embankment side has dropped several centimetres, manifested in displaced paving slabs, cracks and a repaired section of road. The open-channel chute spillway is designed to cope with floodwaters. The energy of floodwater is reduced by the lip at the bottom of the spillway and by the drop to the bottom of the valley. A gated or an open spillway intake could have been used, but the latter was chosen because gates may jam during an earthquake.

Within the Pakhna rocks exposed on both sides of the dam there are excellent sedimentary features, such as slumps and filled channels. Well below the control building there are good examples of the intersection of bedding and cleavage. On the slopes on the western side of the dam, unstable rockfaces are partially terraced, netted and coated with sprayed concrete (shotcrete) to reduce rockfall. However, these measures do not prevent marl flowing onto the road during periods of heavy rain. The view northwards shows that the lower valley sides of the reservoir have been cut back to approximately 45° to increase slope stability.

Stop 7.11 The Palaichori–Kampi dam along the F912 east of Apliki From the E903 a short distance south of Apliki, take the F912 heading southeast towards Farmakas, Kampi and Machairas. Soon the road climbs, giving a good view of the dam, and there is parking just below the dam before the road cuts sharply back to the left. An information board gives details of the dam, reservoir and the uses of water (36 511 557 E, 38 65 290 N).

The concrete gravity dam (height 33 m, length 132 m) was completed in 1973 and, along with its reservoir (maximum capacity 620 000 m³), occupies the bottom of the relatively narrow steep-sided Kampi Valley, which cuts through sheeted dykes of the Troodos ophiolite (Fig. 7.9b). A dam such as this resists the pressure of the water in the reservoir mainly through its own weight anchoring it into the somewhat constricted valley at the site. The dykes are massive, with blocky jointing and some local shearing and shattering, and they strike subparallel to the length of the dam and dip vertically or steeply up stream. These dykes are favourable for the abutments and floor of the dam, but the small basal area of this gravity dam means that it imposes very high stresses on its foundations, unlike an embankment dam, which has a much larger basal area.

Stop 7.12 Water resources, mine drainage and sulphide deposits northwest of Kalavasos This stop visits localities that consider water and mineral resources in and around the Kalavasos reservoir and areas down stream from it. From junction 15 on the A1 motorway, head northwest on the E106 into Kalavasos. At the southern end of the village there is one of the old ore trains that used to run between the Kalavasos mine and Vasiliko on the coast. Cars may travel through the pretty village, but larger vehicles must take the bypass to the northeast. Just over 1 km beyond the village, the road forks sharply and it is necessary to keep left to follow the southwestern side of the Vasilikos River up to the Kalavasos dam. Drive over the crest of the dam and park at its northern end, which provides a good overview (36 524045 E, 38 51351 N).

The Kalavasos reservoir is located within the Limassol Forest of the Troodos ophiolite and it lies behind a rockfill embankment dam that was completed in 1985. The reservoir has a capacity of 17 million m^3 and it supplies water by pipeline to farms and villages down stream. The dam is built on a foundation of volcanic rocks, with much of the fill coming from local sources, as evidenced by the quarries that can still be seen in slopes around the reservoir.

Looking away from the reservoir, some of the hills are made up of near-horizontal white carbonate rocks of the Lefkara Formation, which rest unconformably on dark volcanic rocks. This relationship is particularly evident on the drive up to the dam from Kalavasos. The view southeast down the valley reveals the former loading area for ore from the Kalavasos mine, the spoil from which is evident to the south and southeast. Of particular note is that much of the vegetation down stream of the loading area appears to be dead, which is possibly an effect of contaminated water draining from the mine.

Travel back across the dam to inspect the rocks at the southern abutment. These are pillow lavas cut by many thin dykes, but check to make sure that none of the dykes are in fact sheetflows. The rocks are moderately to steeply dipping, indicating that they have been rotated. Clearly, this movement occurred before deposition of the much younger near-horizontal Lefkara Formation.

Continue southwards on the road for about 350 m and park on the outside of a very tight bend to the left. Three routes originate here and it is necessary to take the middle one, which is a track heading south to an open pit about 500 m away. Initial outcrops contain pink- and green-tinged pillow lavas that are near vertical, again indicating that they have been rotated. In fact the volcanic rocks and their sulphide deposits in this area are believed to lie in the hanging-wall block of a major extensional detachment fault within the Limassol Forest, and it is this fault that caused the rotation.

The pit is large and impressive, but keep well away from its unstable steep slopes. The northern face exposes a zone of mineralization underlain by volcanic rocks, in which pillow structures are evident and intruded by an apparently shallow-dipping dyke with chilled margins. The mineralization appears to be capped by unmineralized lavas, a relationship that may have also been observed during the walk in from the road. However, remember that these rocks have been rotated. The southern face of the pit contains chalk and marl washed down from higher outcrops of the Lefkara Formation capping the volcanic sequence here.

The open pit is just one of many workings in the area. A total of 13 ore-bodies have been exploited in the Kalavasos district, which was one of the major mineralized areas on the island. The ancients worked the orebodies underground, as did much of the mining activity between 1937 and 1977. It was only towards the end of mining in the area that some bodies were worked by open-pit methods. The massive ore had a grade of 45–51 per cent sulphur and 0.3–3.0 per cent copper, and modern mining alone removed 6 million tonnes of ore.

Return to the parking area and continue down hill towards Kalavasos for about 600 m to a turnoff onto a dirt road on the left at the end of the first major bend around to the right. Take the dirt road onto a large unvege-tated area on the right and find the stream flowing out of a disused mine adit (36 524393 E, 38 50962 N). The stream is mildly acidic and is precipi-tating rust-colour minerals dominated by iron oxides. At certain times the water is covered by an oily film, which is probably a product of microbial activity. All of these features are typical of drainage from sulphide mines. Where the stream enters the Vasilikos River, the vegetation appears to be dead, which may be an indication that the waters are contaminated. Contamination is extensive, as evidenced by the rust-colour precipitates occurring all the way down stream to just north of Kalavasos, where they end abruptly as another stream enters the Vasilikos River. Precipitation of iron oxides and associated minerals is likely to be aided by the alkaline composition of unmineralized volcanic rocks and Lefkara carbonate rocks, and waters draining from them.

Before the dam and water pipeline were constructed, local farmers had to rely heavily on private boreholes and wells in the Quaternary deposits of the river bed. The groundwater they extracted was allegedly contami-nated, which is perhaps not surprising considering the quality of the water draining into the river from the mine. There is also concern that the reser-voir is being contaminated by drainage from the mine, especially as sedi-ments in the reservoir contain highly elevated levels of elements found in mine drainage.

The main aquifers in the area around Kalavasos village are the Quater-nary gravels, sands and silts along the Vasilikos River and adjacent coastal zone, the gypsum of the Kalavasos Formation and the massive chalk of the Lefkara Formation. The latter yields the best-quality groundwater, which is used for both irrigation and drinking. The other two sources are used mainly for irrigation, because dissolution of evaporite minerals leads to water with elevated levels of sulphate and chloride, and water from the Quaternary deposits contains high levels of chloride because of contami-nation by sea water.

Waste and landfill

Solid, liquid and gaseous wastes are inevitable byproducts of domestic, commercial and industrial activities, and Cyprus has a history of waste production that goes back millennia. However, it is waste produced in the past century that is most obvious, mainly heaps of mine spoil and tailings that are good examples of open dumping and are considered elsewhere (stops 2.2, 2.3, 2.4, 3.15, 7.3, 7.4, 7.12 and 7.18). Here the focus is on domestic waste, much of which should enter landfill, although some is still fly-tipped.

The solid and liquid wastes generated in Cyprus that are disposed of in landfill are dumped into natural and artificial depressions. Strictly, liquid waste should be processed at a wastewater treatment plant and only solid waste should enter a landfill in a carefully monitored and controlled manner. This solid waste, much of it domestic, is a heterogeneous mixture capable of reacting with water and air to produce landfill liquid (leachate) and gas. Leachate is rich in organic matter, dissolved mineral salts and bacteria, and has the potential to contaminate land and water if it is not properly contained and managed, particularly in Cyprus where water resources are already highly vulnerable. Effective landfills work on the principle of concentrate and contain, whereby waste is encapsulated in an impermeable cell that does not let water in and prevents the uncontrolled escape of leachate and gas. Impermeability is ensured through the use of clay and artificial liners and cappings, and ideally the landfill is sited on stable rock, sediment, or soil of low permeability. The topography and drainage around a landfill must be such that, should contaminants escape, their subsequent dispersion is minimal. It is left to the reader to decide whether or not the two examples of landfill described below work on the principle of concentrate and contain.

Stop 7.13 The landfill site northeast of Kotsiatis   Kotsiatis is about 17 km south of the centre of Lefkosia and the landfill site is situated to the northeast of the village, where it lies at the end of a paved road that runs northwards for just over 1 km from the E102 between Kotsiatis and junction 7 of the A1 motorway. The turnoff from the E102, marked by rubbish, lies immediately east of a large house. There is parking space close to the site entrance (36 532853 E, 38 75986 N), but it is not possible to park or turn a large coach here, so these should park on the E102. Caution must be exercised because of the many heavy vehicles on the roads.

The landfill occupies a disused quarry that lies within chalks and marls of the Pakhna Formation and marls and sands of the Nicosia Formation. The latter dominate the landscape to the north, and surface drainage in the area is to the north, away from the landfill. Although marl horizons in the Nicosia Formation are generally impermeable, many of the other lithologies are porous and permeable. The site appears to receive mainly domestic waste that, once dumped, is bulldozed into marl excavated from the edge of the site.

Stop 7.14 The landfill site on the F628 north of Agia Marinouda This wastedisposal site lies about 3.5 km east of the outskirts of Pafos and it is reached from the B6 at the eastern end of Geroskipou. From here, head east and then north-northeast on the F628, travelling through the village of Agia Marinouda and passing the church of Agia Marina, and park on a large flattened area of calcareous marl on the right near the crest of a hill, just after taking the right fork in the road. Do not go beyond this point, because from here the road descends into the landfill site. Walk to the eastern edge of the flattened land to obtain an excellent overview of the area (Fig. 7.10; 36 452416 E, 38 47299 N).

The lower slopes and bottom of the river valley are made up of brown Mamonia mudstones, whereas the upper slopes are composed of Lefkara chalks and marls. According to the 1:50 000 map of the area, the waste dump occupies the former sites of a small quarry and a reservoir. Liquid waste is discharged from tankers into the valley bottom and it accumulates where there are abundant reeds and tall grasses. Brown mudstone is dug

Figure 7.10 Landfill site in a valley below an embankment of the A6 motorway north of Agia Marinouda (stop 7.14). Liquid waste is discharged in the valley bottom and solid waste is dumped on the valley sides. Both types of waste are covered with Mamonia mudstone quarried on site just below the motorway embankment.

out of the surrounding valley sides and dumped onto the liquid effluent, presumably to enable clays to absorb the liquid. Solid waste is dumped at a slightly higher level and is eventually covered by a layer of mudstone. The lobe of high ground immediately to the north of the viewpoint is artificial, being made of interlayered solid waste and mudstone. This mudstone is eroding to expose waste, which is then free to move.

Geological hazards

Throughout history, the people of Cyprus have been affected by many natural hazards, such as earthquakes, changes in relative sea level, mass movements, storms, floods and droughts. On the island there are good field exposures showing the effects of mass movements and there is archaeological evidence of earthquakes and related damaging events. These are the two geological hazards considered here, but they manifest themselves in very different ways. Damaging earthquakes originate deep below ground and are relatively rare events (low frequency), but they release huge amounts of energy (high magnitude) over a large area. By contrast, mass movements are relatively frequent surface events that usually involve much less energy and affect a small area.

Earthquakes

Cyprus is situated within a wide zone of convergence between the huge tectonic plates of Africa and Eurasia, and one in which some of the most active fault systems on Earth give rise to relatively frequent and devastating earthquakes, such as those recurring along the north Anatolian fault in Turkey (see Fig. 1.1). Cyprus at present lies within a relatively inactive sector of the zone of convergence, but, nonetheless, it experiences significant seismic activity concentrated along the southern part of the island and extending off shore (Fig. 7.11). Much of this activity is attributable to active faulting and subduction beneath Cyprus associated with the Cyprean arc located to the south of the island, and is evident from the cluster of earthquake foci offshore southwest Cyprus. Another distinct cluster lies north of Lemesos and this is probably related to active faulting in the Yerasa fold and thrust belt. From this distribution of seismic activity, it is not surprising that the majority of earthquakes in the past have had greatest impact on the southern, western and southwestern parts of the island. These areas host a significant proportion of clay-rich lithologies and may receive relatively high levels of precipitation, both of which may exacerbate slope failure and mass movement when accompanied by groundshaking.

In the past there have been times when Cyprus appears to have been

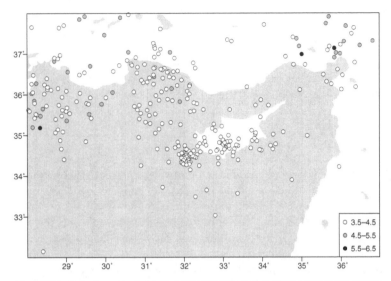

Figure 7.11 Seismic events of magnitude 3.5 and greater in the northeast corner of the eastern Mediterranean for the period April 1997 to January 2004 (data kindly supplied by the Cyprus Geological Survey Department).

much more seismically active than it has been over the past few hundred years, as suggested by archaeological evidence (see Table 7.1). Nevertheless, damaging earthquakes still occur and there were significant events in 1953, 1995 and 1996, which affected the west and southwest of the island. The earthquake of 1953 resulted in the relocation of many villages, and the two events in the 1990s, with epicentres near Polis (1995) and off shore from Pafos (1996), caused a total of 3 deaths, 34 injuries and significant structural damage. Evidently, Cypriots need to understand and manage the risk posed by earthquakes. First and foremost there is a need for careful land-use planning, to avoid problem soils, sediments and rocks, and a requirement for earthquake-resistant structures. Buildings that are one or two storeys tall, constructed of reinforced concrete and founded on solid rock, have the greatest structural integrity during groundshaking.

One of the most effective ways of reducing loss, especially of life, is to be better prepared for an earthquake and to anticipate when the next one will occur. In practice this is very difficult to achieve, but for Cyprus it is currently estimated that a potentially damaging earthquake occurs once every 12 years, a destructive one every 25 years, and a highly destructive one every 120–250 years. It has also been recognized that earthquakes often follow one another very closely in time along a fault or within a larger tectonically active zone. These earthquake clusters or storms probably arise from re-adjustment of the stress field following one initiator earthquake, which then triggers a subsequent earthquake, and so on, such that a seismic chain reaction is established. One such storm may have occurred throughout the Mediterranean in the fourth century AD and it could explain the cataclysmic or universal seismic event of AD 365. This universal event was thought to be the largest documented in the history of the Mediterranean and it caused widespread damage and death in many countries, including Cyprus. However, recent seismic and archae-ological studies have demonstrated that there was not one giant event in AD 365, but 11 large earthquakes over a period of 12 years. One of the last was in AD 370 and this affected southwestern Cyprus. In light of these new findings, the date of the major event that affected Cyprus is now disputed, but here it will be taken as AD 365, with the caveat that it may have occurred five years later. Regardless of when it occurred, this earthquake had an impact on almost every settlement occupied in Cyprus at the time.

Mass movement

Mass movement arises from failure of the land surface brought about by various factors, such as rainfall, loading of slopes, undercutting of bases of slopes, de-vegetation, weathering, earthquakes and fluctuations in the water table. There are five main types of mass movement that develop according to the shape and gradient of the slope and the failure behaviour of the surface material. Massive rigid blocks and slabs of rock normally fail in a brittle manner by falling from steep slopes or sliding down shallower slopes. By contrast, soils, clay-rich lithologies and unconsolidated sedi-ments behave plastically, particularly when they are water saturated, and may move by slow creep or more rapid flow, both of which involve shear-ing throughout the failing mass. Slumps exhibit both brittle and plastic behaviour, and result from rotational slope failures bounded by a spoon-shape basal shear surface.

Most mass movements in Cyprus are associated with weak, easily eroded, clay-rich materials, which are mainly the Mamonia mudstones, the clays of the Kannaviou, Kathikas and Moni Formations, and the marls of the Lefkara, Pakhna and Nicosia Formations (see Figs 6.3, 6.14a, 7.12). The most pronounced and spectacular examples of mass movement are restricted mainly to the southwest of Cyprus, because this is where abun-dant clay-rich lithologies, relief, high rainfall and seismic activity all inter-act. Rockfall is another notable type of slope failure and this occurs most commonly in the ophiolitic rocks of the Troodos hills and mountains.

Where potentially hazardous slopes are encountered or created during development, there is a need for remediation through engineering. There are several options available, most of which are clearly exemplified in the cuttings along the A6 motorway, but please do not stop to examine them.

Figure 7.12 Failure of the E606 because of downslope movement of underlying Mamonia and Kannaviou clays (2 km west of Nata: 36 458724 E, 38 48878 N).

These cuttings are well drained, with cut-off drains along the tops and mid-sections of slopes to catch and deflect surface runoff. At the base of a slope there is a ditch, and sometimes a fence, that serve to channel water and to catch material coming off the slope. It is very noticeable that gradients of cut slopes change with rock type. In the west there are many shallow slopes where the motorway traverses the Kannaviou and Kathikas clays and Mamonia mudstones, because these lithologies are liable to slump, flow and creep. Some parts of these slopes have also been shotcreted or vegetated to prevent erosion, especially of the toes of slopes. In other places, and particularly in the east, it is no coincidence that slopes are much steeper where the motorway traverses rocks mainly of the Pakhna Formation, which are relatively rigid and more prone to fall and slide, except for marl-rich horizons. These steep slopes have been stabilized in various ways: high slopes have been cut back into a series of drained benches that are wide enough to catch most rockfall; faults and joints have been grouted, and sections prone to rapid erosion have been shotcreted; rock bolts and rock anchors have been used to secure slabs of loose or fractured rock; entire slopes have been faced with concrete walls; stepped concrete blocks have been emplaced at the base of cuts to shore up overlying rock; and mesh nets have been used to catch rockfall.

Evidence for hazardous processes

In addition to the archaeological and geological stops described here, there are several other relevant stops. Stops 5.6, 5.8, 6.1, 6.8, 6.14 and 7.10 all provide very good evidence for movement associated with clay-rich lithologies, and stop 2.8 preserves the aftermath of a major rockfall. At stop 7.9 there is excellent evidence for both elevation and submergence of ancient structures as a consequence of tectonic activity.

Stop 7.15 Earthquake evidence, water resources, construction materials and Pakhna and Quaternary deposits at the Kourion ancient site west of Episkopi
This stop is one of the most impressive in Cyprus. It consists of a short excursion that considers geology and archaeology at the Kourion ancient site, which extends 3 km approximately east–west along a narrow plateau overlooking Episkopi Bay. The excursion begins in the west, with an examination of the Sanctuary of Apollon Ylatis, and concludes with an investigation of the archaeology and geology in and around the main site of Kourion. The site is open all year and there is a small entrance fee. The archaeological museum of Kourion is located near by in Episkopi.

Kourion is one of the most important ancient sites in the eastern Mediterranean and is well worth visiting for its remarkable buildings, mosaics and amphitheatre. The entire site comprises the ancient city and early Christian basilica to the east, the Sanctuary of Apollon Ylatis to the west, and the stadium between the two. The city occupies a commanding position on a bluff, with steep cliffs on three sides, and was occupied from Mycenaean times between the fourteenth and twelfth centuries BC. It flourished during the first millennium AD and most of the remains visible today date from Roman times. Although the city was destroyed by earthquakes in the fourth century AD, it was rebuilt early in the fifth century AD and it existed as an early Christian city until it was again destroyed in the seventh century AD, this time during the Arab raids.

The Sanctuary of Apollon Ylatis (Apollo of the Woodlands) lies 3–4 km west of Episkopi, on the northern side of the B6, from where it is well signposted. There is ample parking adjacent to the site entrance (36 487533 E, 38 36686 N). The sanctuary is part of Kourion, but it lies outside the ancient city walls. It was one of the largest and most important religious centres in Cyprus and was used continuously from the late eighth century BC until it was destroyed by the cataclysmic earthquake of AD 365. The restored part of the Roman temple represents the remains of the third temple built on the site, all of them having been destroyed by earthquakes. The restoration is clear to see from the differences in weathering of the building stone (Fig. 7.13), and it involved complicated engineering in an attempt to make the structure resistant to future earthquakes without causing

Figure 7.13 Restored part of the Roman temple at the Sanctuary of Apollon Ylatis (stop 7.15).

aesthetic damage. To do this, reinforced concrete beams and metal bars and dowels were ingeniously used to strengthen the structure internally without leaving a trace on the outside. In some cases the interiors of building stones were cut out and filled with reinforced concrete; the foundations were also strengthened by injection of cement.

Return to the B6 and travel 3 km east before turning right onto Kourion (Curium) Theatre Road East, at the end of which lies the main site of the ancient city of Kourion, with ample parking (36 489780 E, 38 35724 N). This site was also severely damaged during earthquakes in the fourth century

AD, and it is unique in that excavations unearthed a Roman house containing the skeletal remains of a young couple and a child, who are thought to have died during the severe event in AD 365. Other victims recovered at Kourion included a teenage girl and a mule found together in the stable, and a labourer discovered in a doorway. Most of the finds relating to the earthquake are housed in the Kourion museum.

A supply of fresh water to the city was extremely important. Early inhabitants of the site probably depended on stored rainwater, as there are no springs on the bluff. However, as the population grew, a complex system of piping was engineered to supply drinking and bathing water. Perennial springs up to 9 km north of Kourion were tapped by two major gravity-fed conduits, 11 km and 22 km long, which followed contours and were placed at the edges of arable fields to avoid being disturbed by ploughing. These conduits, with regularly spaced settling tanks to collect silt and lime, were made from terracotta pipes laid in earth trenches, in ditches cut into bedrock, or in blocks of stone (that can be seen at the stadium). The conduits delivered water to the fountain house, a public fountain for citizens to draw water, and to the Roman nymphaeum, the main water-supply building dedicated to the nymphs (the protectors of water). From here the rooms in the public baths were fed. There are many other kinds of waterworks at Kourion, dating from the late fourth century BC to the mid-seventh century AD. These include nine deep rock-cut storage cisterns, a large complex of channels and tunnels, channelled limestone blocks and terracotta pipes. These were all probably used to take water to the houses and other buildings in the city.

The main archaeological attractions of the site are undoubtedly the amphitheatre, with its stunning position and view, and the adjacent House of Eustolios and its mosaics. The amphitheatre is Roman, having been remodelled from a smaller Hellenistic circular theatre that existed on the same site, and has been restored recently. These buildings and those on the rest of the site were mainly built from stone locally quarried from the Pakhna Formation. However, rocks from outside Cyprus have been used in the construction of columns in the agora and early Christian basilica, which are made of marble and xenolith-bearing granite, and marble flooring is present in the public baths. The granite was probably imported from western Anatolia and the different-coloured marbles may also be from Turkey. The importation of these exotic rocks, expensive to transport and hard to cut and finish, would have demonstrated the wealth and status of the city.

The view southeast from the amphitheatre emphasizes the commanding position of the city and provides a good overview of the western side of the Akrotiri Peninsula, the western edge of which is a tombolo. This feature probably dates initially from the Pleistocene and it begins at the base of the cliff a short distance west of Kourion (seen at the last locality), and then extends as an arc-shape gravel and sand barrier to Cape Zevgari at the southwestern extremity of the peninsula. In the foreground the land adjacent to the tombolo is made of fertile flat Quaternary deposits, some of which can be examined at stop 6.15 near by. Farther to the southeast is the Akrotiri salt lake (stop 6.18).

For those interested in examining some of the sedimentary rocks of the area, start at the cliffs of Pakhna rocks along the road immediately outside the entrance to the main archaeological site. The smoothness of the cliff faces, with the uniform claw-tooth chisel markings and stonemasons' pick marks, reflect the extensive quarrying that took place at this site. The quarrying provided both building materials and defence measures for the ancient city of Kourion. There are several tombs cut into the flat surface in front of the cliff and into the cliff itself. Starting at the tombs, it is possible to work up the cliff section through a well bedded sequence of carbonate rocks. The quarried levels are mainly within harder calcarenite, with softer chalks above. The lowest bedding planes in the outcrop hosting the tombs contain abundant trace fossils. Overlying these beds is a clastic horizon containing large rounded intraclasts associated with, and overlain by, bioclastic chalk with broken bivalve, gastropod and coral remains. There are also well defined scour marks trending west-southwest–east-northeast. Up section, there appear to be very few, if any, prominent bioclastic horizons. Throughout the sequence, the prominent honeycomb-weathered beds are heavily bioturbated (see Fig. 6.7c).

Now travel down to the beach, where the road bends sharply around to the right and runs northwest across a flat area of sediment between the cliff and the tombolo. Follow this road past restaurants and the remains of another early Christian basilica, and park beyond the last restaurant (36 488714 E, 38 36020 N). The main cliff to the northwest shows several large-scale sedimentary features in the Pakhna Formation. About half-way up the cliff, there is a large and wide channel filled with hard, probably bioclastic, limestones. Below this is a unit of apparently chaotic rocks, with house-size blocks of bedded chalk showing a variety of different dip directions and angles. Some harder, more massive, blocks are pieces of reef limestone. This lower unit appears to represent the top (the base is not seen) of a large slumped horizon, with large blocks having moved down slope. Overlying bedded chalks appear to drape over the top of this slump and can be seen to onlap it. Small bluffs in the less well exposed upper part of the cliff appear to be reef limestones, but it is not clear whether they are *in situ* or transported blocks.

Stop 7.16 Earthquake evidence and building stone at the Pafos archaeological site
The Pafos archaeological site, a UNESCO World Heritage Site, occupies the western part of Kato Pafos and lies immediately north of the modernized harbour (36 445864 E, 38 46152 N). There is a small fee to pay at the entrance to the site, which is located at the western end of a large car-park adjacent to the harbour. The site hosts the famous Roman mosaics in the houses of Dionysos, Orpheus, Theseus and Aion, and also an agora, an odeion (a building for musical and dramatic performances) and the fortress of Saranta Kolones.

The sprawling archaeological site is more correctly known as Nea Pafos (New Pafos), which refers to the ancient city of Pafos, founded in the late fourth century BC. Old Pafos (Palaia Pafos) was located to the southeast, at Kouklia (stop 6.16). Nea Pafos was at its zenith in the second and third centuries AD, but went into decline in the fourth century AD following excessive damage caused by some highly destructive earthquakes. As a consequence, the new capital of Cyprus became Salamis, north of Ammochostos. The situation in Nea Pafos worsened in the seventh century AD because of repeated Arab raids.

The fort of Saranta Kolones was built between AD 1198 and 1204, at the beginning of the Lusignan Period. Saranta Kolones means "40 columns" in Greek, implying that the original structure, built in about AD 1200 on top of its Byzantine predecessor, must have been very grand. However, soon after it was completed it was destroyed by a major earthquake in AD 1222. The arches standing today are not original and have been reconstructed since this seismic event (Fig. 7.14). Before excavations between 1957 and 1983, the only objects clearly visible above ground were the many toppled columns, which can still be seen today. These are made either of xenolith-bearing foliated granodiorite, which is a rock not found in Cyprus and most probably imported from western Anatolia, or of bioclastic calcarenite and limestone, which appear to be the main rock types used in building the fort. To the north of the lighthouse within the archaeological site there are areas of bioclastic calcarenite that have been quarried. These quarries, along with those beside the main road between Kato Pafos and Pafos, and those on the Akamas coast (stop 5.9), are the likely sources of the calcarenite and limestone used in Nea Pafos.

Stop 7.17 Slope failure on the F622 north of Pentalia   This stop is located about 20 km northeast of Pafos, along the F622 between the turnoffs to Pentalia (south) and the F624 to Galataria (north). On this stretch, the F622 runs along the western side of a prominent hill trending north–south and capped by Lefkara chalk. Park close to where the road has failed, or has been repaired, at the base of a steep scarp slope (36 464947 E, 38 57966 N).

Figure 7.14 Massive piers and restored arches at Saranta Kolones (stop 7.16). The piers supported vaults, which were destroyed in the earthquake of AD 1222.

The steep scarred slopes of the hillside are suggestive of past failure of the Lefkara chalks, probably through rockfall. The driving force for this failure would have been plastic flow and erosion of the underlying Kathikas clays and Mamonia mudstones lower down the slope. The road at this point is built on Pleistocene chalk talus, which comprises angular fragments of chalk, mainly chert rich, in a marl matrix. The road has failed by extension and subsidence, and there are large tension gashes, with downthrow on the downslope side of the road. Downthrown and rotated blocks of road dip gently into the hillside as a result of movement of the underlying foundation, further evidence for which are the large arcuate cracks that can be traced from the road down slope into the talus. Clearly, the talus is failing through slumping and flow, but this is magnified because it rests on Kathikas clay and Mamonia mudstone. The failure is probably compounded because of insufficient preparation and compaction of the underlying road foundation.

The hills in the surrounding area are also prone to mass movement and there is plenty of evidence of this preserved. To the north lie the almost abandoned villages of Statos and Agios Fotios, which were badly damaged by earthquakes and mass movement in the twentieth century. Residents were relocated to the new village of Statos–Agios Fotios near by, a situation very similar to that at Choletria (stop 5.6). A little farther north of Statos–Agios Fotios is the religiously very important Panagia Chrysorrogiatissa monastery, which had to be stabilized following the damage it sustained during the 1953 earthquake and subsequent slope movements.

Farther afield, about 10 km to the northwest, there are large failures in the Kannaviou and Kathikas Formations (stop 6.1), and to the south, beyond the turnoff to Nata, a section of the E606 is failing where it passes over Mamonia and Kannaviou clays (see Fig. 7.12).

Stop 7.18 Mining history and waste stabilization at the Amiantos asbestos mine
The mine is located south of the B9 near Pano Amiantos, high up in the Troodos, and it is impossible to miss the huge scar on the landscape. There are many viewpoints, but the best is on the outside of a bend along the B9 to the northwest of Pano Amiantos, at an obvious layby that has a very comprehensive information board (36 492055 E, 38 65564 N). Before considering the mined area, it is informative to examine the cuttings and slopes along the B9 in order to appreciate the highly altered, smashed and unstable nature of the serpentinite in this area. With patience, veins of chrysotile asbestos and picrolite should be found, although better examples exist at stop 2.4.

Mining at Amiantos may date back to ancient times, because the place probably takes its name from "amanthius", which means undefiled and was the name given to asbestos by the ancients. Modern mining of chrysotile asbestos took place between 1904 and 1988, ending mainly because of the slump in the market caused by health concerns. Ore was extracted on a series of open benches on the mountainside, but few, if any, of these remain visible today. The average chrysotile grade was about 0.8–1 per cent, and ore was processed by crushing, hammering, milling, screening, fibring, aspiration and grading into long or short fibres. About half of the ore mined at Amiantos was processed on site, as was apparent from the plume of fine dust that would drift down the valley. Annual production peaked at around 40 000 tonnes of fibre in the late 1940s and early 1950s. As 99 per cent of the ore was unmineralized, this peak production would have generated about 4 million tonnes of spoil each year. In total, a million tonnes of chrysotile asbestos was recovered during the entire operation of the mine, which required the excavation and processing of some 130 million tonnes of rock. The legacy of these activities is the scarred landscape and the huge quantity of spoil and tailings (Fig. 7.15), although this does not represent the total volume of waste produced.

During the first few decades of operation, waste was dumped down slope to be washed away during the winter months. In 1934 this practice caused a huge flood in Kato Amiantos, which killed ten people and prompted a move to constructing stable waste piles on site. Such was the volume of waste generated that it was necessary to infill the Loumata Valley on the southern side of the mine site. In order to permit safe drainage through the waste pile, a 1.3 km-long culvert was constructed along the

Figure 7.15 Aspects of stabilization and failure of the waste pile at the Amiantos asbestos mine (stop 7.18): **(a)** middle and lower slopes of waste in 1987, the penultimate year of mining; **(b)** the same area in 2002 following reprofiling of the slopes, many of which are covered with dead vegetation to reduce erosion; **(c)** reprofiled and partially revegetated slopes above steep barren eroding (note deep gullies) and failing (note slump scar on the far side) lower slopes in 2002.

base of the valley. When the culvert was buried to a depth of 100 m, it structurally failed in places. When this was recognized in 1978, it raised great concern over the stability of the waste pile, and dumping ceased. Today, water that accumulates in the reservoir at the western end of the waste pile seeps slowly through the pile and issues at the base as springs, much to the benefit of farmers down stream who use the water to prolong irrigation during the summer months.

The waste pile remains hazardous today and threatens the safety of Kato Amiantos below (stop 2.4). From Pano Amiantos there are very good views of the steep, barren and permeable lower slopes of the waste pile, which show evidence of movement by creep and slumping (a major slump scar is clearly visible on the southern side of the pile), and erosion by gullying (Fig. 7.15c). Liquefaction is another problem that may result from shaking during an earthquake, but the waste piles are not predicted to fail even with an earthquake of Richter magnitude 7.5 beneath the site. As shown at stop 2.4, the movement of fine sediment that includes chrysotile fibres threatens water quality, and fibres have been reported in fluvial sediments as far down stream as the Trimiklini and Kouris reservoirs.

In 1994 the mine was closed down permanently in order to permit reclamation and redevelopment of the area, which commenced in 1995 and is expected to be complete by 2010. Restoration focuses both on stabilization of spoil, tailings and slopes by reprofiling and revegetating, and on improvement of drainage to transport water to the edges of the waste pile and to reduce erosion and transport of sediment, especially that bearing asbestos fibres. Following terracing of the slopes and attainment of a slope gradient of 2 m horizontal for every 1 m vertical, a ditch is cut into the terrace and is filled with manure and locally derived gabbroic soil, a million cubic metres of which will be needed in total. The soil is then planted or seeded in order to create mixed forest and plant communities similar to those that exist in the surrounding area. An important part of the revegetation and stabilization of the slopes will involve hydro-seeding, whereby a mixture of pulped or mulched organic material, binding substance and seeds will be sprayed onto the slopes to provide a basis for seed germination and growth to form a protective layer against slope erosion. The fully reclaimed site will be used for recreation, tourism and environmental education.

Chapter 8

Excursions

Introduction

All of the individual stops are located in Figure I.1 and are described in Chapters 2 to 7 in a way that permits readers to visit the sites of most interest to them by constructing their own itineraries. However, many readers may wish to follow prescribed itineraries and, therefore, this chapter presents day-long excursions that follow a logical route and examine stops in one or more of the themes adopted in the guide. These itineraries do not allow for much time to be spent at each stop; those travelling with a large group may need to take more time and adjust the excursions accordingly. The information for each excursion merely provides directions from one stop to another. For details on finding exact location, parking and what to see, refer to the specific stop described in a previous chapter.

The Troodos massif and adjacent areas

Excursion A: the Troodos ophiolite from Pano Platres to Mathiatis (see Fig. I.1b) The aim of this excursion is to provide a comprehensive introduction to the main rock types and some of their contact relationships in the Troodos ophiolite. In so doing, the excursion passes through superb landscapes and delightful mountain villages. As far as possible the stops work up through the ophiolite lithostratigraphy, from west to east across the Troodos massif. The full excursion makes a very long day (at least eight hours), but it is well worth the effort.

The excursion begins at an arched bridge on a section of old road on the outside of a bend on the B8, about 2.6 km southwest of Troodos and nearly 5 km north of Pano Platres (second locality of stop 2.7). Here, harzburgite and dunite of the mantle sequence are exposed, but farther up the B8 there are some fine exposures of layered gabbroic rocks of the plutonic sequence (first locality of stop 2.7). Continue up into Troodos, perhaps stopping briefly at the Troodos National Forest Park visitor centre (stop 2.1), and then take the B9 as far as a complex of old white buildings just before the sharp bend to the left on the eastern limit of Pano Amiantos (stop 2.5). The ridge running away from the outside of this bend provides not only exposures of a banded and intruded section of the transition zone, but also excellent views of the Amiantos asbestos mine, details of which are given in stop 7.18.

Continue a short distance down hill from Pano Amiantos, to a major road junction, and turn right onto the E909 and travel eastwards through Kyperounta into Chandria. Here, take the F915 towards Polystypos, to a place 200–300 m beyond the last sharp bend above Chandria, and examine multiple intrusive relationships and water resources in the uppermost plutonic sequence (stop 2.9). Now travel eastwards on the F915, stopping first at the intersection with the E907, close to which there are multiple generations of gabbro exposed (stop 2.10), and then, several kilometres farther on, by two wooden pylons close to a bend between Fterikoudi and Askas. From here there is a view to the north of a body of gabbro and diorite intruded into sheeted dykes (stop 2.12).

From stop 2.12 follow the F915 into Palaichori (keeping to the western side of this village), turn left onto the E903 towards Apliki and, almost immediately, left again onto the E931 into Palaichori. Stop just beyond the first café on the right to study near-vertical sheeted dykes (stop 3.1). Now return to the E903, follow it through Apliki and past the turnoff to Agios Epifanios, and turn right onto the F962 towards Klirou. Stop shortly after, just beyond the bridge over the Akaki River, to view a spectacular section of volcanic rocks and dykes (stop 3.10). Continue to Klirou and then travel through Malounta to Arediou, and take the E907 across badland terrain (stop 6.14) to Episkopeio. From here, travel south through Pera to Kampia on the E902, then east to Analiontas and southeast to Mathiatis through Kataliontas. From Mathiatis take the E103 for about 2 km towards Agia Varvara and park on a large flat barren area on the right, from where sulphide mineralization, umber and volcanic rocks can be examined in and around the Mathiatis mine (stop 3.15).

Excursion B: the Solea graben from Linou to Lemithou (see Figs I.1b and 3.3)
The Solea graben in the Troodos ophiolite provides an opportunity to understand the complex interplay between magmatic and hydrothermal activity, sulphide mineralization and faulting at a submarine spreading centre. Most of the excursion is already detailed in Chapter 3 (stops 3.5 to 3.9; Fig. 3.3), but it could include a visit to the Skouriotissa mine (stop 7.3) if it is open to the public. The excursion begins on the E908, 2.8 km west of its junction with the B9 Lefkosia–Troodos road and about 300 m east of the eastern crossroads for Linou and Katydata, where dykes cutting lavas can

be examined in roadcuts (stop 3.5). From here the mine lies a short distance to the north beyond Katydata (stop 7.3). To continue from stop 3.5, travel 1.8 km west along the E908 to examine tilted lavas and dykes (stop 3.6), and then another 11 km to see listric normal faults and epidotized sheeted dykes (stop 3.7). Continue south to Oikos and take the E911 to 2 km west of Gerakies, where sheeted dykes and epidosites are exposed (stop 3.8). Now travel south to the E912 and follow it either westwards to the monastery of Panagia tou Kykkou, or southeastwards to the F810. Upon joining the F810, travel south to a T-junction and turn left to follow a road into the western side of Lemithou, where it is necessary to stop just below the first houses of the village in order to explore an excellent detachment fault between gabbro and sheeted dykes (stop 3.9).

Excursion C: the Limassol Forest and the Arakapas Valley from Akrounta to Vavla (see Fig. I.1c)
The main aim of this excursion is to provide a comprehensive overview of seafloor spreading, transform and extensional faulting, and magmatism during the formation of the Troodos ophiolite, and it does this by examining the Limassol Forest and adjacent Arakapas Valley, a palaeo-transform fault. However, the excursion begins by considering an effect of structural reorganization of the Limassol Forest during its compression and uplift in the Middle Miocene.

Begin by leaving the A1 motorway at junction 24 and driving north on the F128 to the northern end of the Germasogeia reservoir, to a cutting on the left of the road about 1 km south of Akrounta (stop 6.11). This place marks the start of a 1 km-long walk north into Akrounta, which examines deformed Lefkara marls and chalks in part of the Yerasa fold and thrust belt. From Akrounta, travel about 1 km north on the F128 before turning left onto a forestry road to Apsiou. Some 250 m along this road is an outcrop of serpentinite-bearing volcaniclastic rock sandwiched between two lava flows (stop 4.5). Return to the F128 and continue north for 4 km to just beyond a long cutting on the right of the road, in order to view outcrops exposing intrusions and shear zones in serpentinite (stop 4.6). Drive another 6 km north towards Arakapas and stop in a wide open space on the right of the road, opposite a roadcut exposing a low-angle extensional fault between sheeted dykes and serpentinite (stop 4.7).

Thus far, the stops have considered rocks and structures related to the Limassol Forest, but the remaining stops work up through rocks of the Arakapas fault zone (see Fig. 4.2). From stop 4.7 take the F128 to the southern side of Arakapas, noting that the village sits at the bottom of the prominent east–west-trending Arakapas Valley. Turn right and travel a short distance east along the new bypass, to a cutting on the right that exposes volcanic rocks that were erupted into the trough of the transform fault (stop 4.1). Continue eastwards on the F129 to Eptagoneia and then on the F124 to Akapnou. The next stop is 300 m north of Akapnou, at a roadcut along the F114 towards Ora, where deformed and metamorphosed mafic rocks of the Arakapas fault zone are exposed (stop 4.2). From here continue into Ora and then follow the F112 to the immediate west of Lageia, where the road cuts through a superb section of sedimentary and volcanic rocks that fill the trough of the Arakapas fault zone (stop 4.3). Now take the F112 through Lageia and follow it to just beyond the first turnoff into Vavla. Here there is a track off to the right that permits observation of a thick sequence of sedimentary rocks of the Perapedhi and Kannaviou Formations that are filling the depression of the Arakapas fault zone, and are overlain by Lefkara chalk (stop 4.4). The excursion finishes here, but if time permits, drive down the F112 to Choirokoitia and examine the uppermost Pakhna Formation (stop 6.10) or the Neolithic settlement (stop 7.5).

Excursion D: geological resources and waste disposal between Troodos and Agios Sozomenos (see Fig. I.1b)
The primary aim of this excursion is to investigate exploitation of geological resources and aspects of waste disposal. Begin on the B9 between Troodos and Pano Amiantos, at the obvious parking area with an information board on the outside of a bend that affords an excellent view over the Amiantos asbestos mine (stop 7.18). From here follow the B9 to the first major road junction north of Pano Amiantos and turn right onto the E909, and follow it eastwards through Kyperounta into Chandria. Here, take the F915 towards Polystypos and then travel north on the E907 to meet the E906 between Agia Marina and Kato Moni. At this junction there are some of the best exposures of volcanic rocks in the Troodos ophiolite (stop 3.13), which may be examined if time allows. Turn right onto the E906 and take it south before turning left to follow the E905 to Mitsero, which should be entered from the easternmost turning into the village. From this entrance, follow the road down and turn right onto the F922 towards Kato Moni. Shortly after leaving Mitsero there are two shallow excavations on the right of the road. Stop here to begin a 4 km-long examination of ancient slag, pillow lava, limestone and gypsum quarries, and sulphidic tailings north of Mitsero (stop 7.4). Large coaches must proceed with extreme caution along the narrow roads used by quarry trucks.

At the end of stop 7.4 travel east on the F908, through Agios Ioannis, and turn left onto the E903. After 2 km, turn right and travel through Anageia and Pano Deftara, and then turn right again to Tseri in Kato Deftara. From Tseri, travel south to Margi and then southeast to Kotsiatis to join the E102 heading eastwards to junction 7 of the A1 motorway. A short distance outside Kotsiatis, turn left immediately after a large house and head north for 1 km to the Kotsiatis landfill site (stop 7.13). A large coach would have to park on the E102. Now travel to junction 7 of the A1 motorway and head north towards Lefkosia, exiting at junction 6 and taking the first right to pass under the motorway. At the crossroads, travel straight over on the E120 and later turn right towards Potamia at the T-junction. After about 1 km, turn left towards Geri at a crossroads and head up hill, keeping to the right fork at the farm, to reach the Agios Sozomenos sand quarry for a pre-arranged tour (stop 7.7).

Southwest Cyprus

Excursion E: the Mamonia rocks and the Mamonia/Troodos suture zone from Petra tou Romiou to Agia Marinouda, via the Diarizos and Xeros Potamos Valleys (see Fig. I.1a)

This excursion not only investigates the main rock types of the Mamonia terrane, and their contacts with one another and with rocks of the Troodos ophiolite, but also considers aspects of natural hazards and waste disposal, and permits views of some of the best scenery in southwest Cyprus. The excursion begins in a layby adjacent to the B6 east of Petra tou Romiou and the tourist pavilion, and affords a spectacular view to the west of two promontories, and limestones, volcanic rocks and mudstones of the Dhiarizos Group, which may then be investigated on the beach (stop 5.1). From here, continue towards Pafos on the B6 and turn inland to follow the F616 up the lower part of the Diarizos Valley, along which there are excellent exposures of sedimentary and volcanic rocks of the Dhiarizos and Ayios Photios Groups, which are investigated between 5.6 km and 12 km from the B6 (stops 5.2 to 5.5). It is possible to continue up this delightful valley and observe contacts between the Mamonia rocks and overlying Kannaviou and Lefkara Formations, but to follow the itinerary it is necessary to travel back down the F616 and to turn right onto the F617, which should then be followed to a parking area on the left of the road at the northern end of Nea Choletria (stop 5.6). This place marks the starting point of a mini-excursion that takes at least two hours and examines the Mamonia mélange, Troodos/Mamonia suture zone, and slope instability on both sides of the Xeros Potamos (see Fig. 5.6). The end point is Nata and note that the Xeros Potamos may not be fordable after heavy rain.

From Nata, travel due west to join the E606, noting the serpentinite thrust over Mamonia mudstone along the way, and follow this road for 2.8 km towards Agia Varvara, to a dirt road on the left, and examine metamorphic rocks of the Mamonia terrane (stop 5.7; see Fig. 5.6). Continue southwest to the B6 and follow it westwards to the eastern end of Geroskipou, here turning right onto the F628 and travelling through Agia Marinouda to a flat area near the crest of a hill, from where there is a good view of a landfill sited on Mamonia mudstone (stop 7.14).

Excursion F: geological hazards from Kouklia to Pafos, via Pano Panagia (see Fig. I.1a)

The main aim of this excursion is to assess the causes and impacts of hazardous geological processes operating in southwest Cyprus, but it also takes in some superb archaeological sites and gypsum deposits. Begin at the Sanctuary of Aphrodite and the Palaia Pafos museum near Kouklia, all well signposted from the B6 southeast of Pafos (stop 6.16). From here travel along the B6 towards Pafos before taking the F616 and then the F617 to a parking area on the left of the road at the northern end of Nea Choletria (stop 5.6). Between here and Nata it is possible to examine aspects of ground instability associated with clay-rich Mamonia and Kannaviou lithologies (stop 5.6, localities b, e, f; see Fig. 5.6). Note that the Xeros Potamos between Choletria and Nata may not be fordable after heavy rain. From Nata, travel due west to join the E606, close to where it has failed to the south, and follow this road (becoming the F622) in a northerly direction to just beyond Eledio, where there is parking on a white concrete road off to the left (stop 6.13). From here to Amargeti there are excellent exposures of gypsum of the Kalavasos Formation and of Pliocene sedimentary rocks, all of which formed in the Polis basin.

From Amargeti, continue northeastwards to investigate a failed section of the F622 at the base of a steep slope of white Lefkara chalks, which is beyond the turnoff for Pentalia, but before the intersection with the F624 to Galataria (stop 7.17). Now head north into Pano Panagia and then west on the E703 to Kannaviou, taking note along the way of the evidence for slope failure, the wineries around Pano Panagia, including that at Panagia Chrysorrogiatissa monastery, and the Kannaviou dam. Follow the E703 out of Kannaviou towards Agios Dimitrianos and stop on the right at the small church on a hill to begin an investigation of lithologies of the Kannaviou and Kathikas Formations and mass movement (stop 6.1). On completion, return to the E703 and follow it westwards to the B7, which should be taken south into Pafos. Here, head south to the harbour for the Pafos archaeological site and the earthquake-damaged fortress of Saranta Kolones (stop 7.16).

Southern Cyprus

Excursion G: the circum-Troodos sedimentary succession from Mandria to the Akrotiri salt lake (see Fig. I.1c)

The aim of this excursion is to permit a thorough examination of the Lefkara and Pakhna Formations and some of the Quaternary deposits of the circum-Troodos sedimentary succession. The excursion begins on the E616 between Mandria and Agios Nikolaos, about 5 km west of Mandria, where a contact between Troodos volcanic rocks and Lower Lefkara marls marks the beginning of a complete section through the Lefkara Formation (stop 6.3). From the end of the section, which is near Agios Nikolaos, take the F615 south towards Arsos for about 1.5 km to reach an extensive road exposure of the lower part of the Pakhna Formation (stop 6.8). Now continue through Arsos and Malia to join the E601 heading southeast. Before long, turn right towards Pachna to begin a short traverse through the Pakhna Formation, which starts about 400 m from the E601 (stop 6.9). Return to the E601 and follow it southeast all the way to the B6, which should then be followed westwards through Episkopi before turning left onto Kourion (Curium) Theatre Road East to begin an investigation of Pakhna and Quaternary rocks below the main archaeological site of Kourion (stop 7.15). Return to the B6 and follow it a short distance east before turning right onto the E602 towards Akrotiri. After 2 km, just before a bridge over the Kouris River, turn left and follow a dirt track 50 m north in order to study Pleistocene deposits that record several different environments (stop 6.15). Return to the E602 and follow it east for 2.5 km, then turn right towards Akrotiri, before turning left onto the E604 towards Lemesos. Follow this road for 6 km before turning right onto a road that leads south to a T-junction at the wall of the port of Lemesos. Turn right here, then left at a stop sign, continuing around the port wall. The road soon becomes a dirt track and should be followed south, parallel to the shore of Akrotiri Bay, to just beyond the Lady's Mile café, from where it is possible to gain a brief overview of the Quaternary environments recorded in and around the Akrotiri salt lake (stop 6.18).

Excursion H: the circum-Troodos sedimentary succession from Parekklisia to Larnaka (see Fig. I.1c)

This excursion provides a comprehensive overview of the circum-Troodos sedimentary succession and begins south of the football stadium in Parekklisia, which is reached by taking the E109 northwards from junction 21 of the A1 motorway (stop 6.2). The area south from Parekklisia allows examination of the top of the Troodos volcanic sequence, the Moni mélange, the Lower and Middle Lefkara Formation, the Koronia reef limestones of the Pakhna Formation, and aspects of the Yerasa fold and thrust belt. On completing this stop, travel south on the E109 and then west on the B1 to the Amathous archaeological site, from where exposures of the Pakhna Formation and Quaternary deposits can be explored along the Amathous circuit (stop 7.9). From here travel eastwards on the B1 for about 15 km and then follow the clear signs to Governor's Beach, which is reached from the car-park of the Kalymnos restaurant (stop 6.6). The coast here exposes parts of the Middle and Upper Lefkara Formation, lying within the Yerasa fold and thrust belt, overlain by a Pleistocene marine terrace. Return to the B1 and continue northeastwards along it, before turning south to Psematismenos and Maroni. On leaving the former (note that its streets are too narrow for a large coach), take the first left turning signposted to Maroni. On entering Maroni, turn onto the first paved road on the left and follow it, even after it becomes a dirt road, to where there are excellent exposures of gypsum of the Kalavasos Formation and reef limestone (stop 6.12). Retrace the route back to the B1 and head north along it for a short distance in order to take the F112 as far as the old church in Choirokoitia, from where it is possible to study the uppermost parts of the Pakhna Formation (stop 6.10). Retrace the route along the F112 and head towards Lefkosia by joining the A1 at junction 14. Soon after, at junction 12, join the A5 towards Larnaka and take it to the first roundabout, from where directions to Larnaka airport should be followed. At the airport roundabout take the A4 towards Larnaka and Lefkosia for a short distance, before parking in the first layby in order to examine Pleistocene rocks and modern salt deposits along the southeastern shore of the salt lake (stop 6.17).

Excursion I: resources and hazards from Kourion to Choirokoitia (see Fig. I.1c)
This excursion primarily considers construction materials, water resources, earthquake activity and sea-level change, and, in so doing, takes in several of the premier archaeological sites in Cyprus. The excursion begins along the B6 to the west of Episkopi, at the well signposted Sanctuary of Apollon Ylatis at the western end of the Kourion ancient site (stop 7.15). Over the entire site, earthquake evidence, water supply and construction materials are considered. From the eastern end of the Kourion site, join the eastbound carriageway of the A6 motorway after taking the B6 through Episkopi to Erimi. Once on the A6, immediately take the exit for Ypsonas and Kouris dam, and then turn right onto the F816 and follow it northwestwards to a point where the dam first comes into view (stop 7.10). Return to the A6 and travel east, the motorway soon becoming the A1, and exit at junction 22. Head south to the B1 coast road and turn left to reach the entrance to the ancient city of Amathous, where construction materials,

water supply and sea-level change can be investigated (stop 7.9). Continue eastwards along the B1 and rejoin the A1 at junction 21. Proceed to junction 15 and take the E106 northwest into Kalavasos, and then continue up the Vasilikos Valley, keeping left where the road forks about 1 km beyond the village, to the Kalavasos dam to consider water and mineral resources and water quality (stop 7.12). Return to the A1, follow it northeast to junction 14, and exit for the F112 and Choirokoitia Neolithic settlement, where building stone and stone implements are examined (stop 7.5).

Selected walks

Cyprus provides many opportunities for walking, and the number of well marked trails on the island is increasing. The two walking excursions presented here are located in the Troodos massif, and both allow a full day to be spent examining excellent geology and scenery. Walking is also recommended on the Akamas and around Cape Gkreko, despite there being very few stops in these areas.

Excursion J: the Troodos mantle sequence around Mount Olympos (see Fig. I.1b)
This excursion permits a thorough investigation of the ultramafic and mafic rocks of the mantle sequence and mantle–crust transition zone of the Troodos ophiolite exposed on the slopes of Mount Olympos. The all-day walk is about 16 km long, starting and finishing at the Troodos National Forest Park visitor centre in Troodos (stop 2.1), where information on the area and the walk should be obtained. The walk follows the Atalante trail (stop 2.2), but involves a detour to the Chrome mine (stop 2.3).

Excursion K: the Troodos sheeted dykes along the Madari Ridge (see Fig. I.1b)
The Madari ridge allows examination of some of the very best exposures of sheeted dykes and epidosites in the Troodos ophiolite, in addition to providing some excellent near and far views. The half- or all-day walk begins and ends at a CYTA station, which is reached from the F915 between Chandria and Polystypos (stop 3.3).

Glossary

It should be noted that some definitions might vary slightly from those that are conventionally accepted, as they are intended to be specific to Cyprus. Some definitions have been adapted from various editions of the *Glossary of geology*, published by the American Geological Institute.

accretionary prism A wedge of rock and sediment that accumulates on the front or underside of the non-subducting plate at a convergent boundary.

acid mine drainage Mine drainage water that becomes acidic as a consequence of the oxidation of sulphide minerals.

agora A market place.

alkali, alkaline basalt A basalt relatively poor in silica and rich in the alkali elements sodium and potassium.

amphibolite A dark metamorphic rock mainly made of amphibole and plagioclase feldspar, which formed under moderate to high pressure and temperature.

amygdaloidal Describes igneous rock that contains vesicles filled with one or more secondary minerals (amygdales).

andesite A dark fine-grain volcanic igneous rock having a composition between basalt and rhyolite, and usually containing phenocrysts of plagioclase feldspar and mafic minerals; it is the fine-grain equivalent of diorite.

aphyric A term applied to an igneous rock of fine grain, which does not contain phenocrysts.

aquiclude A body of relatively impermeable material that does not yield useful volumes of water.

aquifer A body of porous and permeable material that stores and transmits useful volumes of water.

armour stone Lumps of hard rock used as protection against the physical action of water, especially waves.

assimilation See **hybridization**.

axial planar cleavage See **cleavage**.

badland Highly unstable land made of clay, silt or marl, sometimes containing soluble minerals, and dissected by many streams.

bastite A serpentine mineral resulting from the alteration of orthopyroxene.

beachrock A sedimentary rock formed in the intertidal zone by precipitation of calcium carbonate, which cements beach sediment.

benthic Pertaining to the bottom of the ocean.

bentonite A pale soft plastic and highly porous sedimentary rock composed of clay minerals of the montmorillonite group and colloidal silica. It forms by the alteration of fine-grain to glassy igneous material, such as volcanic ash.

bioclastic A sedimentary rock mainly composed of fragments of organisms, such as shells.

bioturbation The process by which unconsolidated sediment is disturbed by the activity of organisms.

black smoker A type of hydrothermal vent on the ocean floor. The black smoke is a plume containing mineral-rich particulates that precipitate from the hot hydrothermal fluid when it meets cold sea water.

boninite A type of andesite enriched in both magnesium and silica.

boudinage A structure common in highly deformed rocks where a layer has been stretched, thinned and broken into elongate sausage-shape segments known as boudins.

bryozoan A tiny colonial organism that builds a stony skeleton resembling that of a coral.

calc-alkaline Pertaining to igneous rocks and activity typically associated with subduction-zone environments, particularly volcanic arcs built on continental crust.

calcarenite A type of sandstone composed of calcareous fragments and quartz grains cemented together by calcium carbonate.

calciturbidite A calcareous turbidite.

calcrete A hard conglomerate made up of surficial sand and gravel cemented together by calcium carbonate.

caliche A hard calcareous layer of secondary origin that forms within or on soils in arid and semi-arid regions.

carbonate compensation depth The depth in the ocean below which calcium carbonate is not normally preserved because the rate of its dissolution exceeds the rate of its deposition.

celadonite A soft mica exhibiting a distinct bluish-green colour.

celestite A mineral composed of strontium sulphate.

chevron fold A kink fold in which the limbs are of equal length.

chilled margin The fine-grain edge of an igneous intrusion that is formed by relatively rapid cooling against country rock.

chromite A black chrome-rich oxide mineral that belongs to the spinel group; a common but usually minor constituent of mafic and ultramafic rocks.

chromitite An extremely dark and dense rock composed almost entirely of chromite.

chrysotile A type of asbestos that is a fibrous and silky white to off-white mineral of the serpentine group.

cleavage Regular breaks or partings in rock, which are normally a consequence of metamorphism or deformation. Closely spaced parallel fractures and joints arising from deformation define fracture cleavage. If the cleavage is either parallel or subparallel to the axial plane of a fold, it is an axial planar cleavage. Flaggy cleavage occurs when rock splits into thin slabs.

clinopyroxene A calcium-rich pyroxene.

colloform A term used to describe a variety of mineral textures, which broadly exhibit a rounded and very finely banded kidney-like form.

colluvium Unconsolidated heterogeneous material that accumulates at the base of a slope.

commandery A landed estate under the control of a commander of an order of knights.

competent Describes a material that is able to support itself and is also strong during deformation.

copepod A type of crustacean.

cross bedding See **cross stratification**.

cross lamination See **cross stratification**.

cross stratification Bedding or lamination inclined at an angle to the main stratification.

cumulate The term applied to igneous rock that formed by the accumulation of crystals that settled out of magma.

current mark Any feature on a surface that results from the action of a current of water.

cut-off trench An excavated and filled trench built beneath the foundation baseline of a dam in order to prevent the escape of reservoir water through groundflow.

cyanobacteria Blue-green algae that obtain their energy from photosynthesis.

debris flow A type of mass movement involving unconsolidated and saturated debris, which often exhibits fluid behaviour and moves rapidly.

debrite A sedimentary deposit arising from a debris flow.

diabase An altered dyke rock of mafic composition.

diachronous A term applied to a rock unit in which similar material varies in age from one place to another or that transgresses time planes.

diapirism The process by which a core of mobile material intrudes upward into brittle rock. The Earth's surface may be uplifted, often to form a dome, and ruptured as a consequence. The intrusion is known as a diapir.

diorite A plutonic igneous rock that has a coarse grain and a composition between that of gabbro and granite. It is pale to dark, depending upon the amounts of darker amphibole and pyroxene and paler plagioclase. When the composition is closer to that of granite than gabbro, the rock is granodiorite. Diorite is the intrusive equivalent of andesite.

dolerite A hard fine- to medium-grain intrusive mafic igneous rock having a composition similar to that of basalt and gabbro.

drag fold A fold that develops in relatively soft rock as a consequence of movement.

ductile The term used to describe a material that is able to sustain considerable deformation through plastic behaviour before fracturing.

dunite A peridotite containing at least 90 per cent olivine.

duplex A term used to describe a structural complex in which a set of overlapping thrust faults are contained between an upper (roof) and a basal (floor) thrust fault.

epidote A distinctive greenish hydrous silicate that is indicative of hydrothermal metamorphism. The process of producing epidote is known as epidotization, and a metamorphic rock consisting of mainly epidote and quartz is an epidosite.

epidosite, epidotization See **epidote**.

evaporite A chemical sedimentary rock formed by the evaporation of a saline solution, such as sea water.

exotic A body of rock of any size that is unrelated to the rocks with which it is associated.

facies The properties and characteristics of a rock unit that reflect the conditions under which it formed.

fanglomerate A sedimentary rock consisting of slightly worn fragments of varying size that were initially deposited in an alluvial fan.

ferromagnesian Pertaining to iron and magnesium.

fissile Capable of splitting into thin layers.

flaggy cleavage See **cleavage**.

flame structure A soft-sediment deformation structure resembling a flame or a wave, and formed by the compression and upward intrusion of fine sediment into an overlying layer.

flute cast A sole structure defined by a bulge on the underside of a sedimentary bed where it fills a scoop-shape depression (a flute) in the underlying bed.

fold and thrust belt A compressional feature defined by fold- and thrust-bearing mountainous foot-hills, which run adjacent to a mountain belt.

foliation A planar deformation feature in a rock, and thus a term used specifically in the description of metamorphic rocks.

footwall The underlying side of fault, which lies beneath the hanging wall.

foram A single-cell organism that does not have organs or tissues, but does have a shell (normally made of calcite), and which usually inhabits the marine environment.

fractional crystallization The process by which the composition of a magma evolves by the successive crystallization of minerals and their subsequent separation from the main body of the remaining melt.

fracture cleavage See **cleavage**.

Gondwana See **Pangaea**.

gossan The iron-rich and intensely weathered zone that caps a sulphide deposit. It contains quartz and an array of hydrated iron oxide and sulphate minerals left behind from the oxidation of sulphides and the leaching out of sulphur and most metals.

graben An elongate depression bounded by normal faults along which a crustal block has been downthrown. A half-graben exists where there is faulting on only one side of the depression.

graded bedding Sedimentary bedding in which the grain size grades from coarse to fine, generally from the base upwards.

granodiorite See **diorite**.

greenschist facies An assemblage of minerals (facies) typical of low-temperature and low-pressure metamorphism of mafic rocks. The green colour derives from the presence of epidote and chlorite.

groove cast A sole structure defined by a ridge on the underside of a sedimentary bed where it fills a groove in the underlying bed.

half-graben See **graben**.

hanging wall The overlying side of a fault, which lies above the footwall.

harzburgite A peridotite composed principally of olivine and orthopyroxene.

hemipelagic deposit Material that accumulates in the marine environment near continents and is, therefore, a mixture of pelagic and terrigenous material.

hopper crystal A cubic salt crystal whose edges have grown more extensively than the faces.

hornblende A very common type of amphibole.

horst An elongate and relatively uplifted block bounded by faults on its long sides.

hyaloclastite A sediment made up of angular fragments of basaltic glass.

hybridization The process by which rocks of compositions different from those that would form from the parent magma alone are produced as the result of the parent magma becoming contaminated by incorporating and digesting (assimilating) pre-existing rock. The contaminated magma and the rock it produces are both said to be hybrid.

hydrometallurgical Relating to the use of aqueous chemical processes to leach, concentrate and recover metal from ores, mineral concentrates and recycled or residual materials.

intraclast A broad term used to identify a fragment of ripped up and reworked sedimentary material enclosed in another sediment or sedimentary rock of the same or similar age.

isoclinal fold A tight fold in which the limbs run parallel.

jarosite See **oxyhydroxysulphate**.

kink fold A fold with straight limbs

and a sharp angular hinge.

lherzolite A peridotite composed principally of olivine, orthopyroxene and clinopyroxene.

lineament An extensive linear topographical feature indicative of a structure in the crust.

lineation A linear feature in a rock, such as aligned mineral grains on a foliation plane in a metamorphic rock.

listric Curved.

load cast, load structure An irregular bulge on the base of a sedimentary bed where it has sunk downwards into a pre-existing depression in underlying softer material.

mafic A term used to describe igneous rocks that are dark and contain abundant ferromagnesian minerals and plagioclase feldspar.

magnesite A mineral composed of magnesium carbonate.

marmara A finely laminated variety of gypsum.

massive sulphide deposit An ore deposit made of metal sulphides (mainly pyrite) that formed during accumulation of sulphide minerals precipitated on or below the sea floor during black-smoker activity.

mélange A unit consisting of a heterogeneous mixture of rock and sediment types, where fragments exhibiting a wide range of sizes, compositions and textures are set in a matrix of finer-grain material.

mesa An isolated mass of elevated land with a characteristic flat top and steep sides.

metasedimentary rock A metamorphosed sedimentary rock.

metavolcanic rock A metamorphosed volcanic rock.

meteoric water Water recently derived from the atmosphere.

microgabbro A gabbro possessing a relatively fine grain, but coarser than dolerite.

Moho A shortened version of the more correct term Mohorovičić discontinuity, which marks the sharp change in seismic-wave velocity observed at the mantle/crust boundary. The Moho may lie above the true mantle/crust contact, which probably lies a little deeper in the transition zone and is referred to as the petrologic Moho by some scientists.

montmorillonite   A group of expansive clays.

nannofossil   An extremely small calcareous microfossil.

Neotethyan, Neotethys   See **Tethys**.

nymphaeum   A monumental fountain.

obduction   The process whereby a portion of oceanic lithosphere is moved upwards and emplaced onto the leading edge of a continental mass.

obsidian   A distinctive black and highly siliceous volcanic glass, which fractures to produce extremely sharp edges and shards.

ochre   A red, brown or yellow pigment composed of various hydrated iron oxides, which may exist as well bedded sediment associated with submarine sulphide deposits.

odeion   A building for musical and dramatic performances.

olistostrome   A heterogeneous and chaotic sedimentary deposit formed by slumping or gravity sliding.

onlap   An overlap characterized by successively younger sedimentary units extending progressively farther across an older rock surface.

ooid   A small round or spheroid accretionary mass in a sedimentary unit; smaller than a pisolith.

ophiolite   A distinctive assemblage of ultramafic and mafic rock types that represents a section of the oceanic crust and underlying upper mantle; ophiolites are thought to have formed originally in sea-floor spreading environments, from where they were subsequently tectonically uplifted and emplaced onto continental margins and island arcs.

orthopyroxene   A calcium-poor pyroxene.

oxyhydroxysulphate   A mineral containing oxygen, hydroxide and sulphate, such as the iron-bearing mineral jarosite.

palagonite   A brown, orange or yellow alteration product of basaltic glass.

Pangaea   A supercontinent made up of all the continents that exist today and which was fully assembled by the end of the Palaeozoic era. Pangaea began to break up in the Mesozoic and this initially produced Gondwana (a group of southern continents) and Laurasia (a group of northern continents).

passive continental margin   A largely tectonically stable continental margin that hosts a thick sequence of flat-lying shallow-water sediments.

patch reef   An isolated reef of limited lateral extent, which exists as a mound, a lens or a flat-top body.

pegmatite   The term used to describe an igneous rock exhibiting an exceptionally coarse grain size.

pelagic   Pertaining to the open ocean and not the ocean floor.

peridotite   A group of ultramafic plutonic rocks composed principally of olivine (at least 40%) and pyroxene (less than 60%).

phyric   A suffix used to describe the phenocrysts present in an igneous rock, such as olivine-phyric to identify the phenocrysts as olivine.

picrolite   A greenish columnar or fibrous variety of antigorite, which belongs to the serpentine group of minerals.

pillow lava   A term describing lava that erupts under water and develops a characteristic pillow or tear-drop shape where it comes to rest on the tops of pre-existing pillows.

pisolith   A pea-like accretionary mass in a sedimentary unit; larger than an ooid.

plagioclase   Sodium–calcium feldspar.

plagiogranite   A plutonic igneous rock, light in colour and composed of plagioclase feldspar and quartz, and lesser amounts of biotite, amphibole, pyroxene and alkali feldspar.

plug   A volcanic landform created where solidified magma in a former volcanic vent is left standing proud following erosion of the surrounding rock.

poikilitic   An igneous texture found in plutonic rocks, in which larger crystals of one mineral type host smaller crystals of another.

pressure solution A process of recrystallization in the presence of fluid, whereby highly stressed parts of minerals are dissolved and then they are either precipitated in regions of low stress or transported by the fluid out of the rock. Insoluble material may be redistributed and concentrated during this process.

pteropod A small pelagic swimming gastropod, also known as a sea butterfly.

pyroxenite A group of ultramafic plutonic rocks composed principally of pyroxene (at least 60%) and olivine (less than 40%).

quasi-dyke A dyke-like body of non-igneous origin.

radiolaria Single-cell pelagic marine micro-organisms that build an intricate siliceous skeleton. Deep marine sedimentary rocks may contain high proportions of these skeletons, as is the case with radiolarian cherts and mudstones.

rodingite A calcium-rich rock formed during the process of serpentinization, and usually replacing mafic rock hosted by serpentinized peridotite.

salina A body of saline water in which the concentration of soluble salts is very high.

sanidine An alkali feldspar.

scaphopod A benthic marine mollusc whose shell resembles the shape of an elephant's tusk, which is why it is also referred to as a tusk shell.

selenite A colourless, clear and crystalline form of gypsum. Crystals that have grown freely often have a swallowtail appearance due to the symmetrical intergrowth of crystals along a central twin plane.

serpentine A group of hydrous ferromagnesian silicate minerals derived during the alteration of magnesium-rich silicate minerals, especially olivine.

serpentinite A rock composed almost entirely of serpentine group minerals, which forms by hydrothermal alteration (serpentinization) of ultramafic rocks.

serpulid A type of segmented worm that builds a contorted tube on a submerged surface.

sheeted dykes Vast expanses of rock made up entirely of multiple tabular dykes and sections of dykes that were produced during repeated dyke-in-dyke intrusion.

sheetflow A laterally extensive lava flow that forms a continuous sheet.

shotcrete Sprayed concrete.

slickenside A smooth and polished striated surface resulting from movement on a fault.

sole structure An irregularity, penetration or directional structure located on the underside of a sedimentary bed.

stockwork A network of veins containing economically viable minerals and often found underlying a massive orebody.

strike-slip fault A fault along which slip is parallel to the strike of the fault.

stromatolite A shallow-water accretionary structure created by the activity of micro-organisms, particularly blue-green algae, and consisting of layers of sediment and precipitates.

supercooling The process of lowering the temperature of a liquid below the point at which it should normally crystallize.

supergene enrichment The near-surface process of mineral dissolution and reprecipitation that is often associated with the formation or enrichment of sulphide deposits.

supra-subduction zone The region above the descending plate in a subduction zone, and generally referring to the island-arc environment.

suture zone See **terrane**.

swallowtail See **selenite**.

sylvite A mineral composed of potassium chloride.

tailings The waste material, usually of fine grain, left after mineral processing.

terra rossa A distinctive red residual soil dominated by iron oxides and found overlying limestone in Mediterranean environments.

terrane A fragment of lithosphere from one tectonic plate that has been accreted (sutured) to the lithospheric component belonging to another plate.

The geology of the accreted fragment is distinct from that of the surrounding area. The edge of a terrane is marked by a suture zone, which usually is defined by a major fault or series of faults.

terrigenous Derived from land.

Tethys An ocean that dominated in Mesozoic time and which originally existed in the region now generally occupied by the Alpine–Himalayan mountain belt. It is thought that there were two main oceans, known as Pal-aeotethys and Neotethys. The Mediter-ranean Sea represents a remnant of Neotethys and formed around the end of the Cretaceous period.

tholeiite, tholeiitic basalt A basalt relatively rich in silica.

tombolo A narrow piece of land, such as a sand and gravel bar or barrier, which connects two islands or an island and the mainland.

trace fossil A feature in a sedimentary rock made by the movement, boring, resting, feeding or burrowing of an organism, and preserved as an imprint or a raised structure.

trachyandesite A volcanic rock with a composition intermediate between tra-chyte and andesite.

trachybasalt A volcanic rock with a composition intermediate between tra-chyte and basalt.

trachyte A fine-grain, often feldspar-phyric, volcanic rock, having alkali feldspar and minor biotite, hornblende or pyroxene as the main constituents.

transform fault A type of strike-slip fault, which usually dips steeply or vertically, and one that offsets mid-ocean ridges to accommodate seafloor spreading.

tube worm A general term for worm-like invertebrates that reside in tubes.

tuff A rock consisting of pyroclastic material, particularly volcanic ash.

turbidite A sedimentary deposit formed from a turbidity current and typically exhibiting graded bedding.

turbidity current A rapidly moving density current, consisting of sediment suspended in water, which flows down slope in an ocean or a lake.

ultramafic Said of igneous and meta-morphosed igneous rocks that are dark and dense, and composed principally of mafic minerals (usually olivine and pyroxene). Peridotites and pyroxenites are the most common ultramafic rocks.

umber A fine-grain submarine sedi-mentary deposit containing iron and manganese oxides derived from the black smoke of hydrothermal venting.

variolitic An igneous texture defined by pea-like masses (varioles), normally of radiating crystals, set in a finer-grain matrix.

varve A single sedimentary unit deposited in still water over the course of a year.

wehrlite A peridotite composed prin-cipally of olivine and clinopyroxene.

xenolith A fragment of rock incorpo-rated into magma during its emplace-ment or eruption.

zeolite A mineral belonging to a large group of white or colourless hydrous aluminosilicates that form under very low-grade metamorphic conditions.

Further information and bibliography

Useful contacts

Cyprus Geological Survey Department
1 Lefkonos Street, 1415 Lefkosia,
Cyprus
Tel.: +357 22 409213
Fax: +357 22 316873
E-mail: director@gsd.moa.gov.cy
Web: www.moa.gov.cy/gsd

Cyprus Tourism Organisation
Central Office
Leoforos Lemesou 19, Aglantzia,
2112 Lefkosia, Cyprus
Tel.: +357 22 691100
Fax: +357 22 331644
E-mail: cytour@visitcyprus.com
Web: www.visitcyprus.com

Department of Lands and Surveys
29 Michalakopoulou Street,
1075 Lefkosia, Cyprus
Tel.: +357 22 804801
Fax: +357 22 804872
E-mail: director@dls.moi.gov.cy
Web: www.moi.gov.cy/dls

Government of Cyprus
Web: www.cyprus.gov.cy
Web: www.aspectsofcyprus.com

Water Development Department
100–110 Kennenty Avenue,
Pallouriotissa, 1047 Lefkosia, Cyprus
Tel.: +357 22 609000
Fax: +357 22 675019
E-mail: eioannou@wdd.moa.gov.cy
Web: www.moa.gov.cy/wdd

Publications

Geological and visitor guidebooks
Dubin, M. 2009. *The Rough Guide to Cyprus*. London: Rough Guides.
Greensmith, T. 1994. *Geologists' Association guide 50: southern Cyprus*. London: The Geologists' Association.
Maric, V. 2009. *Cyprus*. Victoria: Lonely Planet Publications.
Xenophontos, C. & J. G. Malpas (eds) 1987. *Field excursion guidebook: Troodos 87 Ophiolites and Oceanic Lithosphere symposium*. Nicosia: Cyprus Geological Survey Department.

Geological maps
Cyprus Geological Survey Department 1970. *1:250000 Hydrogeological map of Cyprus*. Nicosia: The Department.
Cyprus Geological Survey Department 1982. *1:250000 Mineral resources map of Cyprus*. Nicosia: The Department.
Cyprus Geological Survey Department 1995. *1:250000 Geological map of Cyprus*. Nicosia: The Department.

Regional context of Cyprus in the eastern Mediterranean
Dilek, Y., P. Thy, E. M. Moores, T. W. Ramsden 1990. Tectonic evolution of the Troodos ophiolite within the Tethyan framework. *Tectonics* **9**, 811–23.
Garfunkel, Z. 1998. Constraints on the origin and history of the eastern Mediterranean basin. *Tectonophysics* **298**, 5–35.
Moores, E. M., P. T. Robinson, J. Malpas, C. Xenophontos 1984. Model for the origin of the Troodos massif, Cyprus, and other mideast ophiolites. *Geology* **12**, 500–503.
Papazachos, B. C., E. E. Papadimitriou,

A. A. Kiratzi, C. B. Papazachos, E. K. Louvari 1998. Fault plane solutions in the Aegean Sea and the surrounding area and their tectonic implication. *Bollettino di Geofisica Teorica ed Applicata* **39**, 199–218.

Robertson, A. H. F. 1998. Mesozoic–Tertiary tectonic evolution of the easternmost Mediterranean area: integration of marine and land evidence. In *Proceedings of the Ocean Drilling Program, Scientific Results 160*, A. H. F. Robertson, K-C. Emeis, C. Richter, A. Camerlenghi (eds), 723–82. Ocean Drilling Program, Texas A&M University, College Station.

—— 1998. Tectonic significance of the Eratosthenes Seamount: a continental fragment in the process of collision with a subduction zone in the eastern Mediterranean (Ocean Drilling Program Leg 160). *Tectonophysics* **298**, 63–82.

—— 2004. Development of concepts concerning the genesis and emplacement of Tethyan ophiolites in the eastern Mediterranean and Oman regions. *Earth-Science Reviews* **66**, 331–87.

Robertson, A. H. F. & D. Mountrakis (eds) 2006. *Tectonic development of the eastern Mediterranean region*. Special Publication 260, Geological Society, London.

Robertson, A. H. F., P. D. Clift, P. J. Degnan, G. Jones 1991. Palaeogeographic and palaeotectonic evolution of the eastern Mediterranean Neotethys. *Palaeogeography, Palaeoclimatology, Palaeoecology* **87**, 289–343.

Robertson, A. H. F. & M. Grasso 1995. Overview of the Late Tertiary–Recent tectonic and palaeo-environmental development of the Mediterranean region. *Terra Nova* **7**, 114–27.

Geology of Cyprus

Robertson, A. H. F. 1990. Tectonic evolution of Cyprus. In *Ophiolites – oceanic crustal analogues: proceedings of the symposium "Troodos 1987"*, J. Malpas, E. M. Moores, A. Panayiotou, C. Xenophontos (eds), 235–50. Nicosia: Cyprus Geological Survey Department.

Robertson, A. & C. Xenophontos 1993. Development of concepts concerning the Troodos ophiolite and adjacent units in Cyprus. In *Magmatic processes and plate tectonics*, H. M. Prichard, T. Alabaster, N. B. W. Harris, C. R. Neary (eds), 85–119. Special Publication 76, Geological Society, London.

Ophiolites and oceanic lithosphere

Cann, J., M. Nuttall, A. Scott, A. Barnicoat, E. McAllister 2000. *Oceanic crust and ophiolites*. Manchester: UK Earth Science Courseware Consortium (www.ukescc.co.uk).

Harrison, C. G. A. & E. Bonatti 1981. The oceanic lithosphere. In *The sea*, volume 7: *the oceanic lithosphere*, C. Emiliani (ed.), 21–48. New York: John Wiley.

Malpas, J. & P. Robinson 1997. Oceanic lithosphere 1. The origin and evolution of oceanic lithosphere: introduction. *Geoscience Canada* **24**, 100–107.

Nicolas, A. 1989. *Structures of ophiolites and dynamics of oceanic lithosphere*. Dordrecht: Kluwer.

Pearce, J. A. 2003. Supra-subduction zone ophiolites: the search for modern analogues. In *Ophiolite concept and the evolution of geological thought*, Y. Dilek & S. Newcomb (eds), 269–93. Special Paper 373, Geological Society of America, Boulder, Colorado.

Troodos ophiolite: overview

Clube, T. M. M., K. M. Creer, A. H. F. Robertson 1985. Palaeorotation of the Troodos microplate, Cyprus. *Nature* **317**, 522–5.

Constantinou, G. 1980. Metallogenesis associated with Troodos ophiolite. In *Ophiolites: proceedings of the International Ophiolite symposium, Cyprus 1979*, A. Panayiotou (ed.), 663–74. Nicosia: Cyprus Geological Survey Department.

Gass, I. G. 1980. The Troodos Massif: its role in the unravelling of the ophiolite problem and its significance in the understanding of constructive plate margin processes. In *Ophiolites: proceedings of the International Ophiolite symposium, Cyprus 1979*, A. Panayiotou (ed.), 23–35. Nicosia: Cyprus Geological Survey Department.

Moores, E. M. & F. J. Vine 1971. The Troodos massif, Cyprus and other ophiolites as oceanic crust: evaluation and implications. *Royal Society of London, Philosophical Transactions* **A268**, 443–66.

Pearce, J. A. 1975. Basalt geochemistry used to investigate past tectonic environments on Cyprus. *Tectonophysics* **25**, 41–67.

Robinson, P. T., J. Malpas, C. Xenophontos 2003. The Troodos massif of Cyprus: its role in the evolution of the ophiolite concept. In *Ophiolite concept and the evolution of geological thought*, Y. Dilek & S. Newcomb (eds), 295–308. Special Paper 373, Geological Society of America, Boulder, Colorado.

Wilson, R. A. M. & F. T. Ingham 1959. *The geology of the Xeros–Troodos area with an account of the mineral resources.* Memoir 1, Cyprus Geological Survey Department, Nicosia.

Troodos ophiolite: mantle and plutonic sequences

Abelson, M., G. Baer, A. Agnon 2001. Evidence from gabbro of the Troodos ophiolite for lateral magma transport along a slow-spreading mid-ocean ridge. *Nature* **409**, 72–5.

Bartholomew, I. D. 1993. The interaction and geometries of diapiric uprise centres along mid-ocean ridges – evidence from mantle fabric studies of ophiolite complexes. In *Magmatic processes and plate tectonics*, H. M. Prichard, T. Alabaster, N. B. W. Harris, C. R. Neary (eds), 245–56. Special Publication 76, Geological Society, London.

Batanova, V. G. & A. V. Sobolev 2000. Compositional heterogeneity in subduction-related mantle peridotites, Troodos massif, Cyprus. *Geology* **28**, 55–8.

Benn, K. & R. Laurent 1987. Intrusive suite documented in the Troodos ophiolite plutonic complex, Cyprus. *Geology* **15**, 821–4.

Borradaile, G. J. & F. Lagroix 2001. Magnetic fabrics reveal upper mantle flow fabrics in the Troodos ophiolite complex, Cyprus. *Journal of Structural Geology* **23**, 1299–317.

Coogan, L. A., G. J. Banks, K. M. Gillis, C. J. MacLeod, J. A. Pearce 2003. Hidden melting signatures recorded in the Troodos ophiolite plutonic suite: evidence for widespread generation of depleted melts and intra-crustal melt aggregation. *Contributions to Mineralogy and Petrology* **144**, 484–505.

Gillis, K. M. & L. A. Coogan 2002. Anatectic migmatites from the roof of an ocean ridge magma chamber. *Journal of Petrology* **43**, 2075–2095.

Granot, R., M. Abelson, H. Ron, A. Agnon 2006. The oceanic crust in 3D: paleomagnetic reconstruction in the Troodos ophiolite gabbro. *Earth and Planetary Science Letters* **251**, 280–92.

Greenbaum, D. 1972. Magmatic processes at ocean ridges: evidence from the Troodos massif, Cyprus. *Nature (Physical Science)* **238**, 18–21.

Greenbaum, D. 1977. The chromitiferous rocks of the Troodos ophiolite complex, Cyprus. *Economic Geology* **72**, 1175–94.

Hébert. R. & R. Laurent 1990. Mineral chemistry of the plutonic section of the Troodos ophiolite: new constraints for genesis of arc-related ophiolites. In *Ophiolites – oceanic crustal analogues: proceedings of the symposium "Troodos 1987"*, J. Malpas, E. M. Moores, A. Panayiotou, C. Xenophontos (eds), 149–63. Nicosia: Cyprus Geological Survey Department.

Laurent, R., C. Dion, Y. Thibault 1991. Structural and petrological features of peridotite intrusions from the Troodos ophiolite, Cyprus. In *Ophiolite genesis and evolution of the oceanic lithosphere*, Tj. Peters, A. Nicolas, R. G. Coleman (eds), 175–94. Dordrecht: Kluwer.

Malpas, J. 1990. Crustal accretionary processes in the Troodos ophiolite, Cyprus: evidence from field mapping and deep crustal drilling. In *Ophiolites – oceanic crustal analogues: proceedings of the symposium "Troodos 1987"*, J. Malpas, E. M. Moores, A. Panayiotou, C. Xenophontos (eds), 65–74. Nicosia: Cyprus Geological Survey Department.

Malpas, J. & T. Brace 1987. *1:10000 Geological map of the Amiandos–Palekhori area.* Nicosia: Cyprus Geological Survey Department.

McElduff, B. & E. F. Stumpfl 1991. The chromite deposits of the Troodos complex, Cyprus – evidence for the role of a fluid phase accompanying chromite formation. *Mineralium Deposita* **26**, 307–318.

Mukasa, S. B. & J. N. Ludden 1987. Uranium–lead isotopic ages of plagiogranites from the Troodos ophiolite, Cyprus, and their tectonic significance. *Geology* **15**, 825–8.

Thy, P. 1987. Magmas and magma chamber evolution, Troodos ophiolite, Cyprus. *Geology* **15**, 316–19.

Thy, P. & Y. Dilek 2003. Development of ophiolitic perspectives on models of oceanic magma chambers beneath active spreading centers. In *Ophiolite concept and the evolution of geological thought*, Y. Dilek & S. Newcomb (eds), 187–226. Special Paper 373, Geological Society of America, Boulder, Colorado.

Troodos ophiolite: sheeted dykes, volcanic rocks and hydrothermal systems

Agar, S. M. & K. D. Klitgord 1995. A mechanism for decoupling within the oceanic lithosphere revealed in the Troodos ophiolite. *Nature* **374**, 232–8.

Allerton, S. & F. J. Vine 1987. Spreading structure of the Troodos ophiolite, Cyprus: some paleomagnetic constraints. *Geology* **15**, 593–7.

—— 1991. Spreading evolution of the Troodos ophiolite, Cyprus. *Geology* **19**, 637–40.

Alt, J. C. 1994. A sulfur isotopic profile through the Troodos ophiolite, Cyprus: primary composition and the effects of seawater hydrothermal alteration. *Geochimica et Cosmochimica Acta* **58**, 1825–40.

Bednarz, U. & H-U. Schmincke 1989. Mass transfer during sub-seafloor alteration of the upper Troodos crust (Cyprus). *Contributions to Mineralogy and Petrology* **102**, 93–101.

—— 1994. Petrological and chemical evolution of the northeastern Troodos extrusive series, Cyprus. *Journal of Petrology* **35**, 489–523.

Bettison-Varga, L., P. Schiffman, D. R. Janecky 1995. Fluid–rock interaction in the hydrothermal upflow zone of the Solea graben, Troodos ophiolite, Cyprus. In *Low-grade metamorphism of mafic rocks*, P. Schiffman & H. W. Day (eds), 81–100. Special Paper 296, Geological Society of America, Boulder, Colorado.

Bickle, M. J., D. A. H. Teagle, J. Beynon, H. J. Chapman 1998. The structure and controls on fluid–rock interactions in ocean ridge hydrothermal systems: constraints from the Troodos ophiolite. In *Modern ocean floor processes and the geological record*, R. A. Mills & K. Harrison (eds), 127–52. Special Publication 148, Geological Society, London.

Borradaile, G. J. & D. Gauthier 2006. Magnetic studies of magma-supply and sea-floor metamorphism: Troodos ophiolite dikes. *Tectonophysics* **418**, 75–92.

Boyle, J. F. & A. H. F. Robertson 1984. Evolving metallogenesis at the Troodos spreading axis. In *Ophiolites and oceanic lithosphere*, I. G. Gass, S. J. Lippard, A. W. Shelton (eds), 169–81. Special Publication 13, Geological Society, London.

Cann, J. 2003. The Troodos ophiolite and the upper ocean crust; a reciprocal traffic in scientific concepts. In *Ophiolite concept and the evolution of geological thought*, Y. Dilek & S. Newcomb (eds), 309–321. Special Paper 373, Geological Society of America, Boulder, Colorado.

Cann, J. & K. Gillis 2004. Hydrothermal insights from the Troodos ophiolite, Cyprus. In *Hydrogeology of the oceanic lithosphere*, E. E. Davis & H. Elderfield (eds), 272–308. Cambridge: Cambridge University Press.

Constantinou, G. & G. J. S. Govett 1972. Genesis of sulphide deposits, ochre and umber of Cyprus. *Institution of Mining and Metallurgy, Transactions B* **81**, 34–46.

Cowan, J. & J. Cann 1988. Supercritical two-phase separation of hydrothermal fluids in the Troodos ophiolite. *Nature* **333**, 259–61.

Dietrich, D. & S. Spencer 1993. Spreading-induced faulting and fracturing of oceanic crust: examples from the sheeted dyke complex of the Troodos ophiolite, Cyprus. In *Magmatic processes and plate tectonics*, H. M. Prichard, T. Alabaster,

N. B. W. Harris, C. R. Neary (eds), 121–39. Special Publication 76, Geological Society, London.

Eddy, C. A., Y. Dilek, S. Hurst, E. M. Moores 1998. Seamount formation and associated caldera complex and hydrothermal mineralization in ancient oceanic crust, Troodos ophiolite (Cyprus). *Tectonophysics* 292, 189–210.

Furnes, H., K. Muehlenbachs, O. Tumyr, T. Torsvik, C. Xenophontos 2001. Biogenic alteration of volcanic glass from the Troodos ophiolite, Cyprus. *Geological Society of London, Journal* 158, 75–82.

Gillis, K. M. 2002. The rootzone of an ancient hydrothermal system exposed in the Troodos ophiolite, Cyprus. *Journal of Geology* 110, 57–74.

Gillis, K. M. & M. D. Roberts 1999. Cracking at the magma–hydrothermal transition: evidence from the Troodos ophiolite, Cyprus. *Earth and Planetary Science Letters* 169, 227–44.

Gillis, K. M. & P. T. Robinson 1990. Patterns and processes of alteration in the lavas and dykes of the Troodos ophiolite, Cyprus. *Journal of Geophysical Research* 95B, 21,523–21,548.

Hall, J. M. & J-S. Yang 1995. Constructional features of Troodos type oceanic crust: relationships between dike density, alteration, magnetization, and orebody distribution and their implications for *in situ* oceanic crust. *Journal of Geophysical Research* 100B, 19,973–19,989.

Humphris, S. E. & J. R. Cann 2000. Constraints on the energy and chemical balances of the modern TAG and ancient Cyprus seafloor sulfide deposits. *Journal of Geophysical Research* 105B, 28,477–28,488.

Hurst, S. D., E. M. Moores, R. J. Varga 1994. Structural and geophysical expression of the Solea graben, Troodos ophiolite, Cyprus. *Tectonics* 13, 139–56.

Kelley, D. S., P. T. Robinson, J. G. Malpas 1992. Processes of brine generation and circulation in the oceanic crust: fluid inclusion evidence from the Troodos ophiolite, Cyprus. *Journal of Geophysical Research* 97B, 9307–9322.

Kidd, R. G. W. & J. R. Cann 1974. Chilling statistics indicate an ocean-floor spreading origin for the Troodos complex, Cyprus. *Earth and Planetary Science Letters* 24, 151–5.

Little, C. T. S., J. R. Cann, R. J. Herrington, M. Morisseau 1999. Late Cretaceous hydrothermal vent communities from the Troodos ophiolite, Cyprus. *Geology* 27, 1027–1030.

Mackenzie, G. D., P. K. H. Maguire, L. A. Coogan, M. A. Khan, M. Eaton, G. Petrides 2006. Geophysical constraints on the crustal architecture of the Troodos ophiolite: results from the IANGASS project. *Geophysical Journal International* 167, 1385–401.

Malpas, J. & G. Langdon 1984. Petrology of the Upper Pillow Lava suite, Troodos ophiolite, Cyprus. In *Ophiolites and oceanic lithosphere*, I. G. Gass, S. J. Lippard, A. W. Shelton (eds), 155–67. Special Publication 13, Geological Society, London.

Portnyagin, M. V., L. V. Danyushevsky, V. S. Kamenetsky 1997. Coexistence of two distinct mantle sources during formation of ophiolites: a case study of primitive pillow-lavas from the lowest part of the volcanic section of the Troodos ophiolite, Cyprus. *Contributions to Mineralogy and Petrology* 128, 287–301.

Prichard, H. M. & G. Maliotis 1998. Gold mineralization associated with low-temperature, off-axis, fluid activity in the Troodos ophiolite, Cyprus. *Geological Society of London, Journal* 155, 223–31.

Rautenschlein, M., G. A. Jenner, J. Hertogen, A. W. Hofmann, R. Kerrich, H-U. Schmincke, W. M. White 1985. Isotopic and trace element composition of volcanic glasses from the Akaki canyon, Cyprus: implications for the origin of the Troodos ophiolite. *Earth and Planetary Science Letters* 75, 369–83.

Richards, H. G., J. R. Cann, J. Jensenius 1989. Mineralogical zonation and metasomatism of the alteration pipes of Cyprus sulfide deposits. *Economic Geology* 84, 91–115.

Richardson, C. J., J. R. Cann, H. G. Richards, J. G. Cowan 1987. Metal-depleted root zones of the Troodos ore-forming hydrothermal systems, Cyprus. *Earth and Planetary Science Letters* **84**, 243–53.

Robinson, P. T., W. G. Melson, T. O'Hearn, H-U. Schmincke 1983. Volcanic glass compositions of the Troodos ophiolite, Cyprus. *Geology* **11**, 400–404.

Schiffman, P. & B. M. Smith 1988. Petrology and oxygen isotope geochemistry of a fossil seawater hydrothermal system within the Solea graben, northern Troodos ophiolite, Cyprus. *Journal of Geophysical Research* **93B**, 4612–24.

Schmincke, H-U. & U. Bednarz 1990. Pillow, sheet flow and breccia flow volcanoes and volcano-tectonic hydrothermal cycles in the Extrusive Series of the northeastern Troodos ophiolite (Cyprus). In *Ophiolites – oceanic crustal analogues: proceedings of the symposium "Troodos 1987"*, J. Malpas, E. M. Moores, A. Panayiotou, C. Xenophontos (eds), 185–206. Nicosia: Cyprus Geological Survey Department.

Schouten, H. & C. R. Denham 2000. Comparison of volcanic construction in the Troodos ophiolite and oceanic crust using paleomagnetic inclinations from Cyprus Crustal Study Project (CCSP) CY-1 and CY-1A and Ocean Drilling Program (ODP) 504B drill cores. In *Ophiolites and oceanic crust: new insights from field studies and the Ocean Drilling Program*, Y. Dilek, E. M. Moores, D. Elthon, A. Nicolas (eds), 181–94. Special Paper 349, Geological Society of America, Boulder, Colorado.

Schouten, H. & P. B. Kelemen 2002. Melt viscosity, temperature and transport processes, Troodos ophiolite, Cyprus. *Earth and Planetary Science Letters* **201**, 337–52.

Staudigel, H., K. Gillis, R. Duncan 1986. K/Ar and Rb/Sr ages of celadonites from the Troodos ophiolite, Cyprus. *Geology* **14**, 72–5.

Staudigel, H., L. Tauxe, J. S. Gee, P. Bogaard, J. Haspels, G. Kale, A. Leenders, P. Meijer, B. Swaak, M. Tuin, M. C. Van Soest, E. A. T. Verdurmen, A. Zevenhuizen 1999. Geochemistry and intrusive directions in sheeted dikes in the Troodos ophiolite: implications for mid-ocean ridge spreading centers. *Geochemistry Geophysics Geosystems* **1**, doi:10.1029/1999GC000001.

Taylor, R. N. & R. W. Nesbitt 1988. Light rare-earth enrichment of supra subduction-zone mantle: evidence from the Troodos ophiolite, Cyprus. *Geology* **16**, 448–51.

Thy, P. & C. Xenophontos 1991. Crystallization orders and phase chemistry of glassy lavas from the pillow sequences, Troodos ophiolite, Cyprus. *Journal of Petrology* **32**, 403–428.

Van Everdingen, D. A. 1995. Fracture characteristics of the sheeted dike complex, Troodos ophiolite, Cyprus: implications for permeability of oceanic crust. *Journal of Geophysical Research* **100B**, 19,957–19,972.

Varga, R. J. 1991. Modes of extension at oceanic spreading centers: evidence from the Solea graben, Troodos ophiolite, Cyprus. *Journal of Structural Geology* **13**, 517–37.

—— 2003. The sheeted dike complex of the Troodos ophiolite and its role in understanding mid-ocean ridge processes. In *Ophiolite concept and the evolution of geological thought*, Y. Dilek & S. Newcomb (eds), 323–36. Special Paper 373, Geological Society of America, Boulder, Colorado.

Varga, R. J. & E. M. Moores 1985. Spreading structure of the Troodos ophiolite, Cyprus. *Geology* **13**, 846–50.

Varga, R. J. & E. M. Moores 1990. Intermittent magmatic spreading and tectonic extension in the Troodos Ophiolite: implications for exploration for black smoker-type ore deposits. In *Ophiolites – oceanic crustal analogues: proceedings of the symposium "Troodos 1987"*, J. Malpas, E. M. Moores, A. Panayiotou, C. Xenophontos (eds), 53–64. Nicosia: Cyprus Geological Survey Department.

Varga, R. J., J. S. Gee, L. Bettison-Varga, R. S. Anderson, C. L. Johnson 1999. Early establishment of seafloor hydrothermal systems during structural extension: paleomagnetic evidence from the Troodos ophiolite, Cyprus. *Earth and Planetary Science Letters* **171**, 221–35.

Varga, R. J., J. S. Gee, H. Staudigel, L. Tauxe 1998. Dike surface lineations as magma flow indicators within the sheeted dike complex of the Troodos ophiolite, Cyprus. *Journal of Geophysical Research* **103B**, 5241–56.

Troodos ophiolite: Arakapas Valley and Limassol Forest

Bonhommet, N., P. Roperch, F. Calza 1988. Paleomagnetic arguments for block rotations along the Arakapas fault (Cyprus). *Geology* **16**, 422–5.

Cann, J. R., H. M. Prichard, J. G. Malpas, C. Xenophontos 2001. Oceanic inside corner detachments of the Limassol Forest area, Troodos ophiolite, Cyprus. *Geological Society of London, Journal* **158**, 757–67.

Gass, I. G., C. J. MacLeod, B. J. Murton, A. Panayiotou, K. O. Simonian, C. Xenophontos 1994. *The geology of the southern Troodos transform fault zone.* Memoir 9, Cyprus Geological Survey Department, Nicosia.

MacLeod, C. J., S. Allerton, I. G. Gass, C. Xenophontos 1990. Structure of a fossil ridge–transform intersection in the Troodos ophiolite. *Nature* **348**, 717–20.

MacLeod, C. J. & B. J. Murton 1993. Structure and tectonic evolution of the southern Troodos transform fault zone, Cyprus. In *Magmatic processes and plate tectonics*, H. M. Prichard, T. Alabaster, N. B. W. Harris, C. R. Neary (eds), 141–76. Special Publication 76, Geological Society, London.

Morris, A., K. M. Creer, A. H. F. Robertson 1990. Palaeomagnetic evidence for clockwise rotations related to dextral shear along the southern Troodos transform fault, Cyprus. *Earth and Planetary Science Letters* **99**, 250–62.

Murton, B. J. 1986. Anomalous oceanic lithosphere formed in a leaky transform fault: evidence from the western Limassol Forest complex, Cyprus. *Geological Society of London, Journal* **143**, 845–54.

Murton, B. J. 1989. Tectonic controls on boninite genesis. In *Magmatism in the ocean basins*, A. D. Saunders & M. J. Norry (eds), 347–77. Special

Publication 42, Geological Society, London.

Simonian, K. O. & I. G. Gass 1978. Arakapas fault belt, Cyprus: a fossil transform fault. *Geological Society of America Bulletin* **89**, 1220–30.

Mamonia terrane and Mamonia/ Troodos suture zone

Bailey, W. R., R. E. Holdsworth, R. E. Swarbrick 2000. Kinematic history of a reactivated oceanic suture: the Mamonia complex suture zone, SW Cyprus. *Geological Society of London, Journal* **157**, 1107–126.

Borradaile, G. J. & K. Lucas 2003. Tectonics of the Akamas and Mamonia ophiolites, western Cyprus: magnetic petrofabrics and paleomagnetism. *Journal of Structural Geology* **25**, 2053–2076.

Chan, G. H-N., J. Malpas, C. Xenophontos, C-H. Lo 2007. Timing of subduction zone metamorphism during the formation and emplacement of Troodos and Baer-Bassit ophiolites: insights from ^{40}Ar–^{39}Ar geochronology. *Geological Magazine* **144**, 797–810.

—— 2008. Magmatism associated with Gondwanaland rifting and Neo-Tethyan oceanic basin development: evidence from the Mamonia Complex, SW Cyprus. *Geological Society of London, Journal* **165**, 699–709.

Lapierre, H. 1971. *1:50000 Geological map of the Polis–Paphos area.* Nicosia: Cyprus Geological Survey Department.

Lapierre, H., D. Bosch, A. Narros, G. H. Mascle, M. Tardy, A. Demant 2007. The Mamonia Complex (SW Cyprus) revisited: remnant of Late Triassic intra-oceanic volcanism along the Tethyan southwestern passive margin. *Geological Magazine* **144**, 1–19.

Malpas, J., T. Calon, G. Squires 1993. The development of a late Cretaceous microplate suture zone in SW Cyprus. In *Magmatic processes and plate tectonics*, H. M. Prichard, T. Alabaster, N. B. W. Harris, C. R. Neary (eds), 177–95. Special Publication 76, Geological Society, London.

Malpas, J. & C. Xenophontos 1999. *1:25000 Geological map of the Ayia Varvara–Pentalia area.* Nicosia: Cyprus

Geological Survey Department.

Malpas, J., C. Xenophontos, D. Williams 1992. The Ayia Varvara Formation of SW Cyprus: a product of complex collisional tectonics. *Tectonophysics* **212**, 193–211.

Morris, A., M. W. Anderson, A. H. F. Robertson 1998. Multiple tectonic rotations and transform tectonism in an intraoceanic suture zone, SW Cyprus. *Tectonophysics* **299**, 229–53.

Robertson, A. H. F. & N. H. Woodcock 1979. Mamonia complex, southwest Cyprus: evolution and emplacement of a Mesozoic continental margin. *Geological Society of America Bulletin* **90**, 651–65.

Swarbrick, R. E. 1980. The Mamonia complex of SW Cyprus: a Mesozoic continental margin and its relationship to the Troodos complex. In *Ophiolites: proceedings of the International Ophiolite symposium, Cyprus 1979*, A. Panayiotou (ed.), 86–92. Nicosia: Cyprus Geological Survey Department.

Swarbrick, R. E. 1993. Sinistral strike-slip and transpressional tectonics in an ancient oceanic setting: the Mamonia complex, southwest Cyprus. *Geological Society of London, Journal* **150**, 381–92.

Circum-Troodos sedimentary succession and uplift and emergence of Cyprus

Calon, T. J., A. E. Aksu, J. Hall 2005. The Oligocene–Recent evolution of the Mesaoria Basin (Cyprus) and its western marine extension, eastern Mediterranean. *Marine Geology* **221**, 95–120.

Eaton, S. & A. Robertson 1993. The Miocene Pakhna Formation, southern Cyprus and its relationship to the Neogene tectonic evolution of the eastern Mediterranean. *Sedimentary Geology* **86**, 273–96.

Follows, E. J. 1992. Patterns of reef sedimentation and diagenesis in the Miocene of Cyprus. *Sedimentary Geology* **79**, 225–53.

Govers, R., P. Meijer, W. Krijgsman 2009. Regional isostatic response to Messinian salinity crisis events. *Tectonophysics* **463**, 109–29.

Kähler, G. & D. A. V. Stow 1998. Turbidites and contourites of the Palaeogene Lefkara Formation, southern Cyprus. *Sedimentary Geology* **115**, 215–31.

Lord, A. R., I. Panayides, E. Urquhart, C. Xenophontos 2000. A biochronostratigraphical framework for the Late Cretaceous–Recent circum-Troodos sedimentary sequence, Cyprus. In *Proceedings of the Third International Conference on the Geology of the Eastern Mediterranean*, I. Panayides, C. Xenophontos, J. Malpas (eds), 289–97. Nicosia: Cyprus Geological Survey Department.

McCallum, J. E. & A. H. F. Robertson 1995. Sedimentology of two fan-delta systems in the Pliocene–Pleistocene of the Mesaoria Basin, Cyprus. *Sedimentary Geology* **98**, 215–44.

Orszag-Sperber, F., J. M. Rouchy, P. Elion 1989. The sedimentary expression of regional tectonic events during the Miocene–Pliocene transition in the southern Cyprus basins. *Geological Magazine* **126**, 291–9.

Payne, A. S. & A. H. F. Robertson 1995. Neogene supra-subduction zone extension in the Polis graben system, west Cyprus. *Geological Society of London, Journal* **152**, 613–28.

Peybernès, B., M-J. Fondecave-Wallez, P. Cugny 2005. Diachronism in the sedimentary cover around the Troodos ophiolitic massif (Cyprus). *Société Géologique de France, Bulletin* **176**, 161–9.

Poole, A. J. & A. H. F. Robertson 1991. Quaternary uplift and sea-level change at an active plate boundary, Cyprus. *Geological Society of London, Journal* **148**, 909–921.

Poole, A. J., G. B. Shimmield, A. H. F. Robertson 1990. Late Quaternary uplift of the Troodos ophiolite, Cyprus: uranium-series dating of Pleistocene coral. *Geology* **18**, 894–7.

Robertson, A. H. F. 1976. Pelagic chalks and calciturbidites from the lower Tertiary of the Troodos massif, Cyprus. *Journal of Sedimentary Petrology* **46**, 1007–1016.

—— 1977. The Moni mélange, Cyprus: an olistostrome formed at a destructive plate margin. *Geological Society of*

London, Journal **133**, 447–66.

—— 1977. The Kannaviou Formation, Cyprus: volcaniclastic sedimentation of a probable late Cretaceous volcanic arc. *Geological Society of London, Journal* **134**, 269–92.

—— 1977. Tertiary uplift history of the Troodos massif, Cyprus. *Geological Society of America Bulletin* **88**, 1763–72.

Robertson, A. H. F., S. Eaton, E. J. Follows, J. E. McCallum 1991. The role of local tectonics versus global sea-level change in the Neogene evolution of the Cyprus active margin. In *Sedimentation, tectonics and eustasy: sea-level changes at active margins*, D. I. M. Macdonald (ed.), 331–69. Special Publication 12, International Association of Sedimentologists, Oxford.

Robertson, A. H. F., S. Eaton, E. J. Follows, A. S. Payne 1995. Depositional processes and basin analysis of Messinian evaporites in Cyprus. *Terra Nova* **7**, 233–53.

Robertson, A. H. F. & J. D. Hudson 1974. Pelagic sediments in the Cretaceous and Tertiary history of the Troodos massif, Cyprus. In *Pelagic sediments: on land and under the sea*, K. J. Hsü & H. C. Jenkyns (eds), 403–436. Special Publication 1, International Association of Sedimentologists, Oxford.

Rouchy, J. M., F. Orszag-Sperber, M-M. Blanc-Valleron, C. Pierre, M. Rivière, N. Combourieu-Nebout, I. Panayides 2001. Paleoenvironmental changes at the Messinian/Pliocene boundary in the eastern Mediterranean (southern Cyprus basins): significance of the Messinian Lago-Mare. *Sedimentary Geology* **145**, 93–117.

Swarbrick, R. E. & M. A. Naylor 1980. The Kathikas mélange, SW Cyprus: late Cretaceous submarine debris flows. *Sedimentology* **27**, 63–78.

Wade, B. S. & P. R. Bown 2006. Calcareous nannofossils in extreme environments: the Messinian salinity crisis, Polemi Basin, Cyprus. *Palaeogeography, Palaeoclimatology, Palaeoecology* **233**, 271–86.

Resources, wastes and hazards

Ambraseys, N. N. & R. D. Adams 1993. Seismicity of the Cyprus region. *Terra Nova* **5**, 85–94.

Aupert, P. 2000. *Guide to Amathus*. Nicosia: Bank of Cyprus Cultural Foundation.

Boronina, A., P. Renard, W. Balderer, A. Christodoulides 2003. Groundwater resources in the Kouris catchment (Cyprus): data analysis and numerical modelling. *Journal of Hydrology* **271**, 130–49.

Boronina, A., W. Balderer, P. Renard, W. Stichler 2005. Study of stable isotopes in the Kouris catchment (Cyprus) for the description of the regional groundwater flow. *Journal of Hydrology* **308**, 214–26.

Butzer, K. W. & S. E. Harris 2007. Geoarchaeological approaches to the environmental history of Cyprus: explication and critical evaluation. *Journal of Archaeological Science* **34**, 1932–52.

Charalambides, A., E. Kyriacou, C. Constantinou, J. Baker, B. van Os, G. Gurnari, A. Shiathas, P. van Dijk, F. van der Meer 1998. *Mining waste management on Cyprus: assessment, strategy development and implementation*. Nicosia: Cyprus Geological Survey Department.

Christou, D. 2001. *Kourion: its monuments and local museum*. Nicosia: Filokipros.

Constantinou, G. 1982. Geological features and ancient exploitation of the cupriferous sulphide orebodies of Cyprus. In *Early metallurgy in Cyprus, 4000–500 BC*, J. D. Muhly, R. Maddin, V. Karageorghis (eds), 13–24. Larnaca: Pierides Foundation.

Cyprus Tourism Organisation 2001. *Cyprus: 10000 years of history and civilization*. Lefkosia: The Organisation.

Daszewski, W. A. & D. Michaelides 1988. *Guide to the Paphos mosaics*. Nicosia: Bank of Cyprus Cultural Foundation.

Fasnacht, W. 1999. Excavations at Agia Varvara–Almyras: a review of twelve years of research. Report, Department of Antiquities Cyprus, 179–84.

Fokaefs, A. & G. A. Papadopoulos 2007. Tsunami hazard in the eastern Mediterranean: strong earthquakes and tsunamis in Cyprus and the Levantine Sea. *Natural Hazards* **40**, 503–26.

Glekas, I. P. 2008. Towards sustainable water resource management: a case study in Limassol Cyprus. *International Journal of Environment and Pollution* **33**, 15–29.

Hudson-Edwards, K. A. & S. J. Edwards 2005. Mineralogical controls on storage of As, Cu, Pb and Zn at the abandoned Mathiatis massive sulphide mine, Cyprus. *Mineralogical Magazine* **69**, 695–706.

Le Brun, A. 1997. *Khirokitia: a Neolithic site*. Nicosia: Bank of Cyprus Cultural Foundation.

Malpas, J. 2000. An Earth system approach to the geology of the eastern Mediterranean. In *Proceedings of the Third International Conference on the Geology of the Eastern Mediterranean*, I. Panayides, C. Xenophontos, J. Malpas (eds), 1–8. Nicosia: Cyprus Geological Survey Department.

Marangou, A. (ed.) 1992. *Cyprus, copper and the sea*. Nicosia: Government of Cyprus.

Milnes, E. & P. Renard 2004. The problem of salt recycling and seawater intrusion in coastal irrigated plains: an example from the Kiti aquifer (southern Cyprus). *Journal of Hydrology* **288**, 327–43.

Neal, C. & P. Shand 2002. Spring and surface water quality of the Cyprus ophiolites. *Hydrology and Earth System Sciences* **6**, 797–817.

Papadimitriou, E. E. & V. G. Karakostas 2006. Earthquake generation in Cyprus revealed by the evolving stress field. *Tectonophysics* **423**, 61–72.

Stiros, S. C. 2001. The AD 365 Crete earthquake and possible seismic clustering during the fourth to sixth centuries AD in the eastern Mediterranean: a review of historical and archaeological data. *Journal of Structural Geology* **23**, 545–62.

Wdowinski, S., Z. Ben-Avraham, R. Arvidsson, G. Ekstrom 2006. Seismotectonics of the Cyprian arc. *Geophysical Journal International* **164**, 176–81.

Williams-Thorpe, O. & P. C. Webb 2002. Provenancing of Roman granite columns in Cyprus using non-destructive field portable methods. Report, Department of Antiquities Cyprus, 340–63.

Wong, C. K. & J. Malpas 2000. A case study of acid mine drainage, Sia, Cyprus: preliminary results. In *Proceedings of the Third International Conference on the Geology of the Eastern Mediterranean*, I. Panayides, C. Xenophontos, J. Malpas (eds), 365–76. Nicosia: Cyprus Geological Survey Department.

Index of localities

Subject index

Page numbers: *italics* indicate illustrations; a "t" suffix indicates tables.